Commercial Pilot Practical Test Standards for Airplane (SEL, MEL, SES, MES)

FLIGHT STANDARDS SERVICE
Washington, DC 20591

NOTE

Material in FAA-S-8081-12B will be effective August 1, 2002. All previous editions of the Commercial Pilot—Airplane Practical Test Standards will be obsolete as of this date.

Record of Changes

Change 1 dated 2/22/08

- IntroductPage 9, changed to reflect 400' instead of 200' as the minimum altitude for simulated engine failures.

Change 2 dated 2/03/09

Introduction

- Page 7, changed to reflect Notice 8000.331 (inactive, but can be found at **http://fsims.avs.faa.gov**), Airplanes Equipped with Retractable Landing Gear, Flaps, and FADEC Meet the Definition of a Complex Airplane.

FOREWORD

The Commercial Pilot—Airplane Practical Test Standards (PTS) book has been published by the Federal Aviation Administration (FAA) to establish the standards for commercial pilot certification practical tests for the airplane category, single-engine land and sea; and multiengine land and sea classes. FAA inspectors and designated pilot examiners shall conduct practical tests in compliance with these standards. Flight instructors and applicants should find these standards helpful during training and when preparing for the practical test.

/s/ 4/23/2002

Joseph K. Tintera, Manager

Regulatory Support Division

Flight Standards Service

CONTENTS

INTRODUCTION

General Information

The Flight Standards Service of the Federal Aviation Administration (FAA) has developed this practical test book as the standard that shall be used by FAA inspectors and designated pilot examiners when conducting commercial pilot—airplane practical tests. Flight instructors are expected to use this book when preparing applicants for practical tests. Applicants should be familiar with this book and refer to these standards during their training.

Information considered directive in nature is described in this practical test book in terms, such as "shall" and "must" indicating the actions are mandatory. Guidance information is described in terms, such as "should" and "may" indicating the actions are desirable or permissive, but not mandatory.

The FAA gratefully acknowledges the valuable assistance provided by many individuals and organizations throughout the aviation community who contributed their time and talent in assisting with the revision of these practical test standards.

This practical test standard may be downloaded from the Regulatory Support Division's, AFS-600, web site at **http://afs600.faa.gov**. Subsequent changes to this standard, in accordance with AC 60-27, Announcement of Availability: Changes to Practical Test Standards, will also be available on AFS-600's web site and then later incorporated into a printed revision.

Practical Test Standards Concept

Title 14 of the Code of Federal Regulations (14 CFR) part 61 specifies the AREAS OF OPERATION in which knowledge and skill must be demonstrated by the applicant before the issuance of a commercial pilot certificate or rating. The CFRs provide the flexibility to permit the FAA to publish practical test standards containing the AREAS OF OPERATION and specific TASKs in which pilot competency shall be demonstrated. The FAA shall revise this book whenever it is determined that changes are needed in the interest of safety. ***Adherence to the provisions of the regulations and the practical test standards is mandatory for the evaluation of commercial pilot applicants.***

Practical Test Book Description

This test book contains the following Commercial Pilot—Airplane Practical Test Standards:

Section 1 Airplane—Single-Engine Land and Sea
Section 2 Airplane—Multiengine Land and Sea

The Commercial Pilot Practical Test Standards—Airplane includes the AREAS OF OPERATION and TASKs for the issuance of an initial commercial pilot certificate and for the addition of category ratings and/or class ratings to that certificate.

Practical Test Standards Description

AREAS OF OPERATION are phases of the practical test arranged in a logical sequence within each standard. They begin with Preflight Preparation and end with Postflight Procedures. The examiner, however, may conduct the practical test in any sequence that will result in a complete and efficient test; ***however, the ground portion of the practical test shall be accomplished before the flight portion.***

TASKs are titles of knowledge areas, flight procedures, or maneuvers appropriate to an AREA OF OPERATION. The abbreviation(s) within parentheses immediately following a TASK refer to the category and/or class aircraft appropriate to that TASK. The meaning of each abbreviation is as follows.

ASEL Airplane—Single-Engine Land
AMEL Airplane—Multiengine Land
ASES Airplane—Single-Engine Sea
AMES Airplane—Multiengine Sea

NOTE: When administering a test based on sections 1 and 2 of this PTS, the TASKs appropriate to the class airplane (ASEL, ASES, AMEL, or AMES) used for the test shall be included in the plan of action. The absence of a class indicates the TASK is for all classes. NOTE is used to emphasize special considerations required in the AREA OF OPERATION or TASK.

REFERENCE identifies the publication(s) that describe(s) the TASK. Descriptions of TASKs are not included in these standards because this information can be found in the current issue of the listed reference. Publications other than those listed may be used for references if their content conveys substantially the same meaning as the referenced publications.

These practical test standards are based on the following reference.

14 CFR part 43	Maintenance, Preventive Maintenance, Rebuilding, and Alteration
14 CFR part 61	Certification: Pilots, Flight Instructors, and Ground Instructors
14 CFR part 91	General Operating and Flight Rules
AC 00-6	Aviation Weather
AC 00-45	Aviation Weather Services
AC 61-23/ FAA-H-8083-25	Pilot's Handbook of Aeronautical Knowledge
AC 61-65	Certification: Pilots and Flight Instructors
AC 61-67	Stall and Spin Awareness Training.
AC 61-84	Role of Preflight Preparation

AC 90-48	Pilots' Role in Collision Avoidance
AC 90-66	Recommended Standard Traffic Patterns and Practices for Aeronautical Operations At Airports Without Operating Control Towers
AC 91-69	Seaplane Safety for FAR Part 91 Operations
AC 120-51	Crew Resource Management Training
FAA-H-8083-1	Aircraft Weight and Balance Handbook
FAA-H-8083-3	Airplane Flying Handbook
FAA-H-8083-15	Instrument Flying Handbook
AIM	Aeronautical Information Manual
AFD	Airport Facility Directory
NOTAMs	Notices to Airmen
Other	Pilot Operating Handbook
	FAA-Approved Flight Manual
	Navigation Charts
	Seaplane Supplement

The Objective lists the elements that must be satisfactorily performed to demonstrate competency in a TASK. The Objective includes:

1. specifically what the applicant should be able to do;
2. conditions under which the TASK is to be performed; and
3. acceptable performance standard s.

Use of the Practical Test Standards Book

The FAA requires that all commercial pilot practical tests be conducted in accordance with the appropriate commercial practical test standards and the policies set forth in the INTRODUCTION. Applicants shall be evaluated in ***ALL*** TASKs included in each AREA OF OPERATION of the appropriate practical test standard, unless otherwise noted.

An applicant, who holds at least a commercial pilot certificate seeking an additional airplane category rating and/or class rating at the commercial pilot level, shall be evaluated in the AREAS OF OPERATION and TASKs listed in the Additional Rating Task Table. At the discretion of the examiner, an evaluation of the applicant's competence in the remaining AREAS OF OPERATION and TASKs may be conducted.

If the applicant holds two or more category or class ratings at least at the commercial level, and the ratings table indicates differing required TASKs, the "least restrictive" entry applies. For example, if "ALL" and "NONE" are indicated for one AREA OF OPERATION, the "NONE" entry applies. If "B" and "B, C" are indicated, the "B" entry applies.

In preparation for each practical test, the examiner shall develop a written "plan of action." The "plan of action" shall include all TASKs in each AREA OF OPERATION, unless noted otherwise. If the elements in one TASK have already been evaluated in another TASK, they need not be repeated. For example, the "plan of action" need not include evaluating the applicant on complying with markings, signals, and clearances at the end of the flight, if that element was sufficiently observed at the beginning of the flight. ***Any TASK selected for evaluation during a practical test shall be evaluated in its entirety.***

The examiner is not required to follow the precise order in which the AREAS OF OPERATION and TASKs appear in this book. The examiner may change the sequence or combine TASKs with similar Objectives to have an orderly and efficient flow of the practical test. For example, Radio Communications and ATC Light Signals may be combined with Traffic Patterns. The examiner's "plan of action" shall include the order and combination of TASKs to be demonstrated by the applicant in a manner that will result in an efficient and valid test.

The examiner is expected to use good judgment in the performance of simulated emergency procedures. The use of the safest means for simulation is expected. Consideration must be given to local conditions, both meteorological and topographical, at the time of the test, as well as the applicant's workload, and the condition of the aircraft used. If the procedure being evaluated would jeopardize safety, it is expected that the applicant will simulate that portion of the maneuver.

Special Emphasis Areas

Examiners shall place special emphasis upon areas of aircraft operations considered critical to flight safety. Among these are:

1. positive aircraft control;
2. positive exchange of the flight controls procedure (who is flying the airplane);
3. stall/spin awareness;
4. collision avoidance;
5. wake turbulence avoidance;
6. Land and Hold Short Operations (LAHSO);
7. runway incursion avoidance;
8. controlled flight into terrain (CFIT);
9. aeronautical decision making (ADM);
10. checklist usage; and
11. other areas deemed appropriate to any phase of the practical test.

Although these areas may not be specifically addressed under each **TASK, they are essential to flight safety and will be evaluated during the** practical test. In all instances, the applicant's actions will relate to the complete situation.

Removal of the "Airplane Multiengine VFR Only" Limitation

The removal of the "Airplane Multiengine VFR Only" limitation, at the commercial pilot certificate level, requires an applicant to satisfactorily perform the following AREAS OF OPERATION and TASKs from the commercial AMEL and AMES PTS in a multiengine airplane that has a manufacturer's published V_{MC} speed.

AREA OF OPERATION X: MULTIENGINE OPERATIONS

TASK C: ENGINE FAILURE DURING FLIGHT (By Reference to Instruments)
TASK D: INSTRUMENT APPROACH—ONE ENGINE INOPERATIVE (By Reference to Instruments)

Removal of the "Limited to Center Thrust" Limitation

The removal of the "Limited to Center Thrust" limitation at the commercial pilot certificate level requires an applicant to satisfactorily perform the following AREAS OF OPERATION and TASKs from the commercial AMEL and AMES PTS in a multiengine airplane that has a manufacturer's published V_{MC} speed.

AREA OF OPERATION I: PREFLIGHT PREPARTATION

TASK H: PRINCIPLES OF FLIGHT-ENGINE INOPERATIVE

AREA OF OPERATION X: EMERGENCY OPERATIONS

TASK B: ENGINE FAILURE DURING TAKEOFF BEFORE VMC (SIMULATED)
TASK C: ENGINE FAILURE AFTER LIFT-OFF (SIMULATED)
TASK D: APPROACH AND LANDING WITH AN INOPERATIVE ENGINE (SIMULATED)

AREA OF OPERATION XI: MULTIENGINE OPERATIONS

TASK A: MANEUVERING WITH ONE ENGINE INOPERATIVE
TASK B: VMC DEMONSTRATION

Commercial Pilot—Airplane Practical Test Prerequisites

An applicant for the Commercial Pilot—Airplane Practical Test is required by 14 CFR part 61 to:

1. be at least 18 years of age;
2. be able to read, speak, write, and understand the English language. If there is a doubt, use AC 60-28, English Language Skill Standards;
3. possess a private pilot certificate with an airplane rating, if a commercial

pilot certificate with an airplane rating is sought, or meet the flight experience required for a private pilot certificate (airplane rating) and pass the private airplane knowledge and practical test;

4. possess an instrument rating (airplane) or the following limitation shall be placed on the commercial pilot certificate: "Carrying passengers in airplanes for hire is prohibited at night or on cross-country flights of more than 50 nautical miles;"
5. have passed the appropriate commercial pilot knowledge test since the beginning of the 24th month before the month in which he or she takes the practical test;
6. have satisfactorily accomplished the required training and obtained the aeronautical experience prescribed;
7. possess at least a current third class medical certificate;
8. have an endorsement from an authorized instructor certifying that the applicant has received and logged training time within 60 days preceding the date of application in preparation for the practical test, and is prepared for the practical test; and
9. also have an endorsement certifying that the applicant has demonstrated satisfactory knowledge of the subject areas in which the applicant was deficient on the airman knowledge test.

Aircraft and Equipment Required for the Practical Test

The commercial pilot—airplane applicant is required by 14 CFR section 61.45, to provide an airworthy, certificated airplane for use during the practical test. This section further requires that the aircraft must:

1. be of U.S., foreign or military registry of the same category, class, and type, if applicable, for the certificate and/or rating for which the applicant is applying;
2. have fully functioning dual controls, except as provided for in 14 CFR section 61.45(c) and (e);
3. be capable of performing all AREAS OF OPERATION appropriate to the rating sought and have no operating limitations which prohibit its use in any of the AREAS OF OPERATION required for the practical test; and
4. be a complex airplane furnished by the applicant, unless the applicant currently holds a commercial pilot certificate with a single-engine or multiengine class rating as appropriate, for the performance of takeoffs, landings, and appropriate emergency procedures. A complex landplane is one having a retractable landing gear, flaps, and controllable propeller. A complex seaplane is one having flaps, floats, and controllable propeller. Airplanes equipped with a full authority digital engine control (FADEC) system are considered to have a controllable propeller.

Use of FAA-Approved Flight Simulator or Flight Training Device

An airman applicant for commercial pilot—airplane certification may be authorized to use an FAA-qualified and approved flight simulator or flight training device, to complete certain flight TASK requirements listed in this practical test standard.

An airman applicant seeking an added rating to a commercial certificate may also use a qualified and approved flight simulator or flight training device to complete the flight task requirements in accordance with Appendix 1 and 2 of these practical test standards. These appendices should be consulted to identify which flight TASKs may be accomplished in an approved flight simulator or flight training device. The level of flight simulator or flight training device required for each maneuver or procedure will also be found in the appropriate appendix. An appropriate class airplane is required to complete the remaining flight TASKs for certification.

When flight TASKs are accomplished in an aircraft, certain TASK elements may be accomplished through "simulated" actions in the interest of safety and practicality, but when accomplished in a flight simulator or flight training device, these same actions would not be "simulated." For example, when in an aircraft, a simulated engine fire may be addressed by retarding the throttle to idle, simulating the shutdown of the engine, simulating the discharge of the fire suppression agent, if applicable, simulating the disconnect of associated electrical, hydraulic, and pneumatics systems, etc. However, when the same emergency condition is addressed in a flight simulator or flight training device, all TASK elements must be accomplished as would be expected under actual circumstances.

Similarly, safety of flight precautions taken in the aircraft for the accomplishment of a specific maneuver or procedure (such as limiting altitude in an approach to stall or setting maximum airspeed for an engine failure expected to result in a rejected takeoff) need not be taken when a flight simulator or flight training device is used.

It is important to understand that whether accomplished in an aircraft, flight simulator or flight training device, all TASKs and elements for each maneuver or procedure shall have the same performance standards applied equally for determination of overall satisfactory performance.

Flight Instructor Responsibility

An appropriately rated flight instructor is responsible for training the commercial pilot applicant to acceptable standards in all subject matter areas, procedures, and maneuvers included in the TASKs within each AREA OF OPERATION in the appropriate commercial pilot practical test standard.

Because of the impact of their teaching activities in developing safe, proficient pilots, flight instructors should exhibit a high level of knowledge, skill, and the ability to impart that knowledge and skill to students.

Throughout the applicant's training, the flight instructor is responsible for emphasizing the performance of effective visual scanning and collision avoidance procedures.

Examiner[1] Responsibility

The examiner conducting the practical test is responsible for determining that the applicant meets the acceptable standards of knowledge and skill of each TASK within the appropriate practical test standard. Since there is no formal division between the "oral" and "skill" portions of the practical test, this becomes an ongoing process throughout the test. Oral questioning, to determine the applicant's knowledge of TASKs and related safety factors, should be used judiciously at all times, especially during the flight portion of the practical test. Examiner's shall test to the greatest extent practicable the applicant's correlative abilities rather than mere rote enumeration of facts throughout the practical test.

If the examiner determines that a TASK is incomplete, or the outcome uncertain, the examiner may require the applicant to repeat that TASK, or portions of that TASK. This provision has been made in the interest of fairness and does not mean that instruction, practice, or the repeating of an unsatisfactory task is permitted during the certification process. When practical, the remaining TASKs of the practical test phase should be completed before repeating the questionable TASK.

On multiengine practical tests where the failure of the most critical engine after lift off is required, the examiner must give consideration to local atmospheric conditions, terrain, and type of aircraft used. However the failure of an engine shall not be simulated until attaining at least V_{SSE}/V_{YSE} and at an altitude not lower than 400 feet AGL.

During simulated engine failures on multiengine practical tests the examiner shall set zero thrust after the applicant has simulated feathering the propeller. The examiner shall require the applicant to demonstrate at least one landing with a simulated feathered propeller with the engine set to zero thrust.

Throughout the flight portion of the practical test, the examiner shall evaluate the applicant's use of visual scanning and collision avoidance procedures.

Satisfactory Performance

Satisfactory performance to meet the requirements for certification is based on the applicant's ability to safely:

1. perform the TASKs specified in the AREAS OF OPERATION for the certificate or rating sought within the approved standards;
2. demonstrate mastery of the aircraft with the successful outcome of each TASK performed never seriously in doubt;

1 The word "examiner" is used throughout the standards to denote either the FAA inspector or FAA designated pilot examiner who conducts an official practical test.

3. demonstrate satisfactory proficiency and competency within the approved standards;
4. demonstrate sound judgment; and
5. demonstrate single-pilot competence if the aircraft is type certificated for single-pilot operations.

Unsatisfactory Performance

The tolerances represent the performance expected in good flying conditions. If, in the judgment of the examiner, the applicant does not meet the standards of performance of any TASK performed, the associated AREA OF OPERATION is failed and therefore, the practical test is failed.

The examiner or applicant may discontinue the test at any time when the failure of an AREA OF OPERATION makes the applicant ineligible for the certificate or rating sought. ***The test may be continued ONLY with the consent of the applicant.*** If the test is discontinued, the applicant is entitled credit for only those AREAS OF OPERATION and their associated TASKs satisfactorily performed. However, during the retest, and at the discretion of the examiner, any TASK may be reevaluated, including those previously passed.

Typical areas of unsatisfactory performance and grounds for disqualification are:

1. Any action or lack of action by the applicant that requires corrective intervention by the examiner to maintain safe flight.
2. Failure to use proper and effective visual scanning techniques to clear the area before and while performing maneuvers.
3. Consistently exceeding tolerances stated in the Objectives.
4. Failure to take prompt corrective action when tolerances are exceeded.

When a notice of disapproval is issued, the examiner shall record the applicant's unsatisfactory performance in terms of the AREA OF OPERATION and specific TASK(s) not meeting the standard appropriate to practical test conducted. The AREA(s) OF OPERATION/TASK(s) not tested and the number of practical test failures shall also be recorded. If the applicant fails the practical test because of a special emphasis area, the Notice of Disapproval shall indicate the associated task. i. e.: AREA OF OPERATION VIII, Maneuvering During Slow Flight, failure to use proper collision avoidance procedures.

Crew Resource Management (CRM)

CRM refers to the effective use of all available resources: human resources, hardware, and information. Human resources include all groups routinely working with the cockpit crew or pilot who are involved with decisions that are required to operate a flight safely. These groups include, but are not limited to dispatchers, cabin crewmembers, maintenance personnel, air traffic controllers, and weather services. CRM is not a single TASK, but a set of competencies that must be evident in all TASKs in this practical test standard as applied to either single pilot operations or crew.

Applicant's Use of Checklists

Throughout the practical test, the applicant is evaluated on the use of an appropriate checklist. Proper use is dependent on the specific TASK being evaluated. The situation may be such that the use of the checklist, while accomplishing elements of an Objective, would be either unsafe or impractical, especially in a single-pilot operation. In this case, a review of the checklist after the elements have been accomplished, would be appropriate. Division of attention and proper visual scanning should be considered when using a checklist.

Use of Distractions During Practical Tests

Numerous studies indicate that many accidents have occurred when the pilot has been distracted during critical phases of flight. To evaluate the applicant's ability to utilize proper control technique while dividing attention both inside and/or outside the cockpit, the examiner shall cause realistic distractions during the flight portion of the practical test to evaluate the applicant's ability to divide attention while maintaining safe flight.

Positive Exchange of Flight Controls

During flight training, there must always be a clear understanding between students and flight instructors of who has control of the aircraft. Prior to flight, a briefing should be conducted that includes the procedure for the exchange of flight controls. A positive three-step process in the exchange of flight controls between pilots is a proven procedure and one that is strongly recommended.

When the instructor wishes the student to take control of the aircraft, he or she will say, "You have the flight controls." The student acknowledges immediately by saying, "I have the flight controls." The flight instructor again says, "You have the flight controls." When control is returned to the instructor, follow the same procedure. A visual check is recommended to verify that the exchange has occurred. There should never by any doubt as to who is flying the aircraft.

Metric Conversion Initiative

To assist pilots in understanding and using the metric measurement system, the practical test standards refer to the metric equivalent of various altitudes throughout. The inclusion of meters is intended to familiarize pilots with its use. The metric altimeter is arranged in 10 meter increments; therefore, when converting from feet to meters, the exact conversion, being too exact for practical purposes, is rounded to the nearest 10 meter increment or even altitude as necessary.

SECTION 1

COMMERCIAL PILOT—AIRPLANE

Single-Engine Land and Single-Engine Sea

CONTENTS

Airplane Single-Engine Land
and
Airplane Single-Engine Sea

ADDITIONAL RATING TASK TABLE

Airplane Single-Engine Land

Addition of an Airplane Single-Engine Land Rating to an existing Commercial Pilot Certificate								
Required TASKs are indicated by either the TASK letter(s) that apply(s) or an indication that all or none of the TASKs must be tested based on the notes in each AREA OF OPERATION.								
COMMERCIAL PILOT RATING(S) HELD								
AREAS OF OPER-ATION	**ASES**	**AMEL**	**AMES**	**RH**	**RG**	**Glider**	**Balloon**	**Airship**
I	F,G	F,G	F,G	F,G	F,G	F,G	F,G	F,G
II	D	NONE	D	A,C,D, F	A,D,F	A,B,C, D,F	A,B,C, D,F	A,B,C, D,F
III	C	NONE	C	B,C	NONE	B,C	B,C	B,C
IV	A,B,C, D,E,F, K	A,B,C, D,E,F, K	A,B,C, D,E,F, K	A,B,C, D,E,F, K,L	A,B,C, D,E,F, K,L	A,B,C, D,E,F, K,L	A,B,C, D,E,F, K,L	A,B,C, D,E,F, K,L
V	NONE	B,C,D	B,C,D	ALL	ALL	ALL	ALL	ALL
VI	NONE	ALL	ALL	ALL	ALL	ALL	ALL	ALL
VII	NONE	NONE	NONE	NONE	NONE	ALL	ALL	NONE
VIII	NONE	NONE	NONE	ALL	ALL	ALL	ALL	ALL
IX	A,B	A,B	A,B	ALL	ALL	ALL	ALL	ALL
X	NONE	NONE	NONE	ALL	ALL	ALL	ALL	ALL
XI	A	NONE	A	A	A	A	A	A

ADDITIONAL RATING TASK TABLE

Airplane Single-Engine Land

COMMERCIAL PILOT RATING(S) HELD								
AREAS OF OPER-ATION	ASEL	AMEL	AMES	RH	RG	Glider	Balloon	Airship
I	F,G,H, I	F,G,H, I	F,G	F,G,H, I	F,G,H, I	F,G,H, I	F,G,H, I	F,G,H, I
II	E	E	E	A,B,C, E,F	A,B,E, F	A,B,C, E,F	A,B,C, E,F	A,B,C, E,F
III	C	C	NONE	B,C	C	B,C	B,C	B,C
IV	A,B,E, F,G,H, I,J	A,B,E, F,G,H, I,J	A,B,E, F,G,H, I,J	A,B,E, F,G,H, I,J,K,L	A,B,E, F,G,H, I,J,K,L	A,B,E, F,G,H, I,J,K,L	A,B,E, F,G,H, I,J,K,L	A,B,E, F,G,H, I,J,K,L
V	NONE	B,C,D	B,C,D	ALL	ALL	ALL	ALL	ALL
VI	NONE	ALL	ALL	ALL	ALL	ALL	ALL	ALL
VII	NONE	NONE	NONE	NONE	NONE	ALL	ALL	NONE
VIII	NONE	NONE	NONE	ALL	ALL	ALL	ALL	ALL
IX	A,B	A,B	A,B	ALL	ALL	ALL	ALL	ALL
X	NONE	NONE	NONE	ALL	ALL	ALL	ALL	ALL
XI	B,C,D	B,C,D	NONE	B,C,D	B,C,D	B,C,D	B,C,D	B,C,D

APPLICANT'S PRACTICAL TEST CHECKLIST

APPOINTMENT WITH EXAMINER:

EXAMINER'S NAME______________________________

LOCATION _____________________________________

DATE/TIME ____________________________________

ACCEPTABLE AIRCR

- □ Aircraft Documents:

  Airworthiness Certificate
  Registration Certificate
  Operating Limitations

- □ Aircraft Maintenance Records:

  Logbook Record of Airworthiness Inspections and AD Compliance

- □ Pilot's Operating Handbook, FAA-Approved

  Airplane Flight Manual

PERSONAL EQUIPMENT

- □ View-Limiting Device Current Aeronautical Charts
- □ Computer and Plotter
- □ Flight Plan Form
- □ Flight Logs
- □ Current AIM, Airport Facility Directory, and Appropriate Publications

PERSONAL RECORDS

- □ Identification—Photo/Signature ID
- □ Pilot Certificate
- □ Current and Appropriate Medical Certificate
- □ Completed FAA Form 8710-1, Airman Certificate and/or Rating Application with Instructor's Signature (If applicable)

- □ Computer Test Report
- □ Pilot Logbook with appropriate Instructor Endorsements
- □ FAA Form 8060-5, Notice of Disapproval (if applicable)
- □ Approved School Graduation Certificate (if applicable)
- □ Examiner's Fee (if applicable)

EXAMINER'S PRACTICAL TEST CHECKLIST

Airplane Single-Engine Land And Airplane Single-Engine Sea

APPLICANT'S NAME______________________________

LOCATION______________________________________

DATE/TIME______________________________________

I. PREFLIGHT PREPARATION

□ A. Certificates and Documents (ASEL and ASES)

□ **B.** Airworthiness Requirements (ASEL and ASES)

□ **C.** Weather Information (ASEL and ASES)

□ **D.** Cross-Country Flight Planning (ASEL and ASES)

□ **E.** National Airspace System (ASEL and ASES)

□ **F.** Performance and Limitations (ASEL and ASES)

□ **G.** Operation of Systems (ASEL and ASES)

□ **H.** Water and Seaplane Characteristics (ASES)

□ **I.** Seaplane Bases, Maritime Rules, and Aids to Marine Navigation (ASES)

□ **J.** Aeromedical Factors (ASEL and ASES)

II. PREFLIGHT PROCEDURES

□ A. Preflight Inspection (ASEL and ASES)

□ B. Cockpit Management (ASEL and ASES)

□ C. Engine Starting (ASEL and ASES)

□ D. Taxiing (ASEL)

□ E. Taxiing and Sailing (ASES)

□ F. Before Takeoff Check (ASEL and ASES)

III. AIRPORT AND SEAPLANE BASE OPERATIONS

□ A. Radio Communications and ATC Light Signals (ASEL and ASES)

□ B. Traffic Patterns (ASEL and ASES)

□ C. Airport/Seaplane Base, Runway, and Taxiway Signs, Markings, and Lighting (ASEL and ASES)

IV. TAKEOFFS, LANDINGS, AND GO-AROUNDS

□ A. Normal and Crosswind Takeoff and Climb (ASEL and ASES)

□ B. Normal and Crosswind Approach and Landing (ASEL and ASES)

□ C. Soft-Field Takeoff and Climb (ASEL)

□ D. Soft-Field Approach and Landing (ASEL)

□ E. Short-Field (Confined Area—ASES) Takeoff and Maximum Performance Climb (ASEL and ASES)

□ F. Short-Field Approach (Confined Area—ASES) and Landing (ASEL and ASES)

□ G. Glassy Water Takeoff and Climb (ASES)

□ H. Glassy Water Approach and Landing (ASES)

□ I. Rough Water Takeoff and Climb (ASES)

□ J. Rough Water Approach and Landing (ASES)

□ K. Power-Off 180° Accuracy Approach and Landing (ASEL and ASES)

□ L. Go-Around/Rejected Landing (ASEL and ASES)

V.PERFORMANCE MANEUVERS

□ A. Steep Turns (ASEL and ASES)

□ B. Steep Spiral (ASEL and ASES)

□ C. Chandelles (ASEL and ASES)

□ D. Lazy Eights (ASEL and ASES)

VI. GROUND REFERENCE MANEUVER

□ Eights on Pylons (ASEL and ASES)

VII. NAVIGATION

□ A. Pilotage and Dead Reckoning (ASEL and ASES)

□ B. Navigation Systems and Radar Services (ASEL and ASES)

□ C. Diversion (ASEL and ASES)

□ D. Lost Procedures (ASEL and ASES)

VIII. SLOW FLIGHT AND STALLS

□ A. Maneuvering During Slow Flight (ASEL and ASES)

□ B. Power-Off Stalls (ASEL and ASES)

□ C. Power-On Stalls (ASEL and ASES)

□ D. Spin Awareness (ASEL and ASES)

IX. EMERGENCY OPERATIONS

□ A. Emergency Approach and Landing (Simulated) (ASEL and ASES)

□ B. Systems and Equipment Malfunctions (ASEL and ASES)

□ C. Emergency Equipment and Survival Gear (ASEL and ASES)

X. HIGH ALTITUDE OPERATIONS

□ A. Supplemental Oxygen (ASEL and ASES)

□ B. Pressurization (ASEL and ASES)

XI. POSTFLIGHT PROCEDURES

□ A. After Landing, Parking, and Securing (ASEL and ASES)

□ B. Anchoring (ASES)

□ C. Docking and Mooring (ASES)

□ D. Ramping/Beaching (ASES)

I. AREA OF OPERATION: PREFLIGHT PREPARATION

NOTE: The examiner shall develop a scenario based on real time weather to evaluate TASKs C and D.

A. TASK: CERTIFICATES AND DOCUMENTS (ASEL and ASES)

REFERENCES: 14 CFR parts 43, 61, 91; FAA-H-8083-3; AC 6123/ FAA-H-8083-25; POH/AFM.

Objective. To determine that the applicant exhibits knowledge of the elements related to certificates and documents by:

1. Explaining—
   a. commercial pilot certificate privileges limitations and recent flight experience requirements.
   b. medical certificate class and duration.
   c. pilot logbook or flight records.
2. Locating and explaining—
   a. airworthiness and registration certificates.
   b. operating limitations, placards, instrument markings, and POH/AFM.
   c. weight and balance data and equipment list.

B. TASK: AIRWORTHINESS REQUIREMENTS (ASEL and ASES)

REFERENCES: 14 CFR part 91; AC 61-23/FAA-H-8083-25.

Objective. To determine that the applicant exhibits knowledge of the elements related to airworthiness requirements by:

1. Explaining–
   a. required instruments and equipment for day/night VFR.
   b. procedures and limitations for determining airworthiness of the airplane with inoperative instruments and equipment with and without an MEL.
   c. requirements and procedures for obtaining a special flight permit.
2. Locating and explaining—
   a. airworthiness directives.
   b. compliance records.
   c. maintenance/inspection requirements.
   d. appropriate record keeping

C. TASK: WEATHER INFORMATION (ASEL and ASES)

REFERENCES: 14 CFR part 91; AC 00-6, AC 00-45, AC 61-23/FAA-H8083-25, AC 61-84; AIM.

Objective. To determine that the applicant:

1. Exhibits knowledge of the elements related to weather information by analyzing weather reports, charts, and forecasts from various sources with emphasis on—

    a. METAR, TAF, and FA.

    b. surface analysis chart.

    c. radar summary chart.

    d. winds and temperature aloft chart.

    e. significant weather prognostic charts.

    f. convective outlook chart.

    g. AWOS, ASOS, and ATIS reports.

2. Makes a competent "go/no-go" decision based on available weather information.

D. TASK: CROSS-COUNTRY FLIGHT PLANNING (ASEL and ASES)

REFERENCES: 14 CFR part 91; AC 61-23/FAA-H-8083-25, AC 61-84; Navigation Charts; A/FD; AIM.

Objective. To determine that the applicant:Exhibits knowledge of the elements related to cross-country flight planning by presenting and explaining a pre-planned VFR cross-country flight, as previously assigned by the examiner. On the day of the practical test, the final flight plan shall be to the first fuel stop, based on maximum allowable passengers, baggage, and/or cargo loads using real time weather.

1. Uses appropriate and current aeronautical charts.
2. Properly identifies airspace, obstructions, and terrain features.
3. Selects easily identifiable en route checkpoints.
4. Selects most favorable altitudes considering weather conditions and equipment capabilities.
5. Computes headings, flight time, and fuel requirements.
6. Selects appropriate navigation system/facilities and communication frequencies.

7. Applies pertinent information from NOTAMs, A/FD, and other flight publications.

8. Completes a navigation log and simulates filing a VFR flight plan.

E. TASK: NATIONAL AIRSPACE SYSTEM (ASEL and ASES)

REFERENCES: 14 CFR part 71, 91; Navigation Charts; AIM.

Objective. To determine that the applicant exhibits knowledge of the elements related to the National Airspace System by explaining:

1. Basic VFR weather minimums—for all classes of airspace.
2. Airspace classes—their operating rules, pilot certification, and airplane equipment requirements for the following—
   a. Class A.
   b. Class B.
   c. Class C.
   d. Class D.
   e. Class E.
   f. Class G.
3. Special use and other airspace areas.

F. TASK: PERFORMANCE AND LIMITATIONS (ASEL and ASES)

REFERENCES: AC 61-23/FAA-H-8083-25; FAA-H-8083-1; AC 61-84, POH/AFM.

Objective. To determine that the applicant:

1. Exhibits knowledge of the elements related to performance and limitations by explaining the use of charts, tables, and data to determine performance and the adverse effects of exceeding limitations.
2. Computes weight and balance. Determines if the computed weight and center of gravity is within the airplane's operating limitations and if the weight and center of gravity will remain within limits during all phases of flight.
3. Demonstrates use of the appropriate performance charts, tables, and data.
4. Describes the effects of atmospheric conditions on the airplane's performance.

G. TASK: OPERATION OF SYSTEMS (ASEL and ASES)

REFERENCES: AC 61-23/FAA-H-8083-25; POH/AFM.

Objective. To determine that the applicant exhibits knowledge of the elements related to the operation of systems on the airplane provided for the practical test, by explaining at least five (5) of the following systems.

1. Primary flight controls and trim.
2. Flaps, leading edge devices, and spoilers.
3. Water rudders (ASES).
4. Power plant and propeller.
5. Landing gear.
6. Fuel, oil, and hydraulic.
7. Electrical.
8. Avionics.
9. Pitot-static, vacuum/pressure and associated flight instruments.
10. Environmental.
11. Deicing and anti-icing.

H. TASK: WATER AND SEAPLANE CHARACTERISTICS (ASES)

REFERENCE: FAA-H-8083-3.

Objective. To determine that the applicant exhibits knowledge of the elements related to water and seaplane characteristics by explaining:

1. The characteristics of a water surface as affected by features, such as—
   a. size and location.
   b. protected and unprotected areas.
   c. surface wind.
   d. direction and strength of water current.
   e. floating and partially submerged debris.
   f. sandbars, islands, and shoals.
   g. vessel traffic, and wakes.
   h. other features peculiar to the area.
2. Float and hull construction, and their effect on seaplane performance.
3. Causes of porpoising and skipping, and the pilot action required to prevent or correct these occurrences.

I. TASK: SEAPLANE BASES, MARITIME RULES, AND AIDS TO MARINE NAVIGATION (ASES)

REFERENCES: FAA-H-8083-3; AIM.

Objective. To determine that the applicant exhibits knowledge of the elements related to seaplane bases, maritime rules, and aids to marine navigation by explaining:

1. How to locate and identify seaplane bases on charts or in directories.
2. Operating restrictions at various bases.
3. Right-of-way, steering, and sailing rules pertinent to seaplane operation.
4. Marine navigation aids, such as buoys, beacons, lights, and sound signals.

J. TASK: AEROMEDICAL FACTORS (ASEL and ASES)

REFERENCES: AC 61-23/FAA-H-8083-25; AIM.

Objective. To determine that the applicant exhibits knowledge of the elements related to aeromedical factors by explaining:

1. The symptoms, causes, effects, and corrective actions of at least four (4) of the following—
   a. hypoxia.
   b. hyperventilation.
   c. middle ear and sinus problems.
   d. spatial disorientation.
   e. motion sickness.
   f. carbon monoxide poisoning.
   g. stress and fatigue.
   h. dehydration.
2. The effects of alcohol, drugs, and over-the-counter medications.
3. The effects of excess nitrogen during scuba dives upon a pilot or passenger in flight.

II. AREA OF OPERATION: PREFLIGHT PROCEDURES

A. TASK: PREFLIGHT INSPECTION (ASEL and ASES)

REFERENCES: FAA-H-8083-3; POH/AFM.

Objective. To determine that the applicant:

1. Exhibits knowledge of the elements related to preflight inspection. This shall include which items must be inspected, the reasons for checking each item, and how to detect possible defects.
2. Inspects the airplane with reference to an appropriate checklist.
3. Verifies that the airplane is in condition for safe flight.

B. TASK: COCKPIT MANAGEMENT (ASEL and ASES)

REFERENCES: FAA-H-8083-3; POH/AFM.

Objective. To determine that the applicant:

1. Exhibits knowledge of the elements related to cockpit management procedures.
2. Ensures all loose items in the cockpit and cabin are secured.
3. Organizes material and equipment in an efficient manner so they are readily available.
4. Briefs occupants on the use of safety belts, shoulder harnesses, doors, and emergency procedures.

C. TASK: ENGINE STARTING (ASEL and ASES)

REFERENCES: FAA-H-8083-3, AC 61-23/FAA-H-8083-25, AC 91-13, AC 91-55; POH/AFM.

Objective. To determine that the applicant:

1. Exhibits knowledge of the elements related to recommended engine starting procedures. This shall include the use of an external power source, hand propping safety, and starting under various atmospheric conditions.
2. Positions the airplane properly considering structures, surface conditions other aircraft, and the safety of nearby persons and property.
3. Utilizes the appropriate checklist for starting procedure.

D. TASK: TAXIING (ASEL)

REFERENCES: FAA-H-8083-3; POH/AFM.

Objective. To determine that the applicant:

1. Exhibits knowledge of the elements related to safe taxi procedures.
2. Performs a brake check immediately after the airplane begins moving.
3. Positions flight controls properly for the existing wind conditions.
4. Controls direction and speed without excessive use of brakes.
5. Complies with airport/taxiway markings, signals, ATC clearances and instructions.
6. Taxies so as to avoid other aircraft and hazards.

E. TASK: TAXIING AND SAILING (ASES)

REFERENCES: FAA-H-8083-3; USCG NAVIGATION RULES, INTERNATIONAL-INLAND; POH/AFM.

Objective. To determine that the applicant:

1. Exhibits knowledge of the elements related to water taxi and sailing procedures.
2. Positions the flight controls properly for the existing wind conditions.
3. Plans and follows the most favorable course while taxi or sailing considering wind, water current, water conditions and maritime regulations.
4. Uses the appropriate idle, plow, or step taxi technique.
5. Uses flight controls, flaps, doors, water rudder, and power correctly so as to follow the desired course while sailing.
6. Prevents and corrects for porpoising and skipping.
7. Avoids other aircraft, vessels, and hazards.
8. Complies with seaplane base signs, signals, and clearances.

F. TASK: BEFORE TAKEOFF CHECK (ASEL and ASES)

REFERENCES: FAA-H-8083-3; POH/AFM.

Objective. To determine that the applicant:

1. Exhibits knowledge of the elements related to the before takeoff check. This shall include the reasons for checking each item and how to detect malfunctions.

2. Positions the airplane properly considering other aircraft/vessels, wind and surface conditions.
3. Divides attention inside and outside the cockpit.
4. Ensures the engine temperatures and pressure are suitable for run-up and takeoff.
5. Accomplishes the before takeoff checklist and ensures the airplane is in safe operating condition.
6. Reviews takeoff performance airspeeds, takeoff distances, departure and emergency procedures.
7. Avoids runway incursion and/or ensures no conflict with traffic prior to taxiing into takeoff position.

III. AREA OF OPERATION: AIRPORT AND SEAPLANE BASE OPERATIONS

A. TASK: RADIO COMMUNICATIONS AND ATC LIGHT SIGNALS (ASEL and ASES)

REFERENCES: 14 CFR part 91; AC 61-23/FAA-H-8083-25; AIM.

Objective. To determine that the applicant:

1. Exhibits knowledge of the elements related to radio communications and ATC light signals.
2. Selects appropriate frequencies.
3. Transmits using recommended phraseology.
4. Acknowledges radio communications and complies with instructions.

B. TASK: TRAFFIC PATTERNS (ASEL and ASES)

REFERENCES: FAA-H-8083-3, AC 61-23/FAA-H-8083-25, AC 90-66; AIM.

Objective. To determine that the applicant:

1. Exhibits knowledge of the elements related to traffic patterns. This shall include procedures at airports with and without operating control towers, prevention of runway incursions, collision avoidance, wake turbulence avoidance, and wind shear.
2. Complies with proper traffic pattern procedures.
3. Maintains proper spacing from other aircraft.
4. Corrects for wind-drift to maintain proper ground track.
5. Maintains orientation with runway/landing area in use.
6. Maintains traffic pattern altitude ±100 feet (30 meters), and appropriate airspeed ±10 knots.

C. TASK: AIRPORT/SEAPLANE BASE, RUNWAY, AND TAXIWAY SIGNS, MARKINGS, AND LIGHTING (ASEL and ASES)

REFERENCES: AC 61-23/FAA-H-8083-25; AIM.

Objective. To determine that the applicant:

1. Exhibits knowledge of the elements related to airport/seaplane base, runway, and taxiway operations with emphasis on runway incursion avoidance.
2. Properly identifies and interprets airport/seaplane base, runway, and taxiway signs, markings, and lighting.

IV. AREA OF OPERATION: TAKEOFFS, LANDINGS, AND GO-AROUNDS

A. TASK: NORMAL AND CROSSWIND TAKEOFF AND CLIMB (ASEL and ASES)

NOTE: If a crosswind condition does not exist, the applicant's knowledge of crosswind elements shall be evaluated through oral testing.

REFERENCES: FAA-H-8083-3; POH/AFM.

Objective. To determine that the applicant:

1. Exhibits knowledge of the elements related to normal and crosswind takeoff, climb operations and rejected takeoff procedures.
2. Positions the flight controls for the existing wind conditions.
3. Clears the area, taxies onto the take off surface and aligns the airplane on the runway center/takeoff path.
4. Retracts the water rudders as appropriate (ASES), and advances the throttle smoothly to takeoff power.
5. Establishes and maintains the most efficient planing/lift off attitude and corrects for porpoising and skipping (ASES).
6. Lifts off at the recommended airspeed, and accelerates to VY.
7. Establishes a pitch attitude that will maintain VY,±5 knots.
8. Retracts the landing gear if appropriate, and flaps after a positive rate of climb is established.
9. Maintains takeoff power and VY ±5 knots to a safe maneuvering altitude.
10. Maintains directional control, proper wind-drift correction throughout the takeoff and climb.
11. Complies with noise abatement procedures.
12. Completes appropriate checklists.

B. TASK: NORMAL AND CROSSWIND APPROACH AND LANDING (ASEL and ASES)

NOTE: If a crosswind condition does not exist, the applicant's knowledge of the crosswind elements shall be evaluated through oral testing.

REFERENCES: FAA-H-8083-3; POH/AFM.

Objective. To determine that the applicant:

1. Exhibits knowledge of the elements related to normal and crosswind approach and landing.
2. Adequately surveys the intended landing area (ASES).
3. Considers the wind conditions, landing surface, obstructions, and selects a suitable touchdown point.
4. Establishes the recommended approach and landing configuration and airspeed and adjusts pitch attitude and power as required.
5. Maintains a stabilized approach and recommended airspeed, or in its absence, not more than 1.3 V_{SO}, ±5 knots, with wind gust factor applied.
6. Makes smooth, timely, and correct control application during the roundout and touchdown.
7. Contacts the water at the proper pitch attitude (ASES).
8. Touches down smoothly at approximate stalling speed (ASEL).
9. Touches down at or within 200 feet (60 meters) beyond a specified point, with no drift, and with the airplane's longitudinal axis aligned with and over the runway center/landing path.
10. Maintains crosswind correction and directional control throughout the approach and landing sequence.
11. Completes appropriate checklist.

C. TASK: SOFT-FIELD TAKEOFF AND CLIMB (ASEL)

REFERENCES: FAA-H-8083-3; POH/AFM.

Objective. To determine that the applicant:

1. Exhibits knowledge of the elements related to a soft-field takeoff and climb.
2. Positions the flight controls for existing conditions and to maximize lift as quickly as possible.
3. Clears the area; taxies onto takeoff surface at a speed consistent with safety without stopping while advancing the throttle smoothly to takeoff power.
4. Establishes and maintains a pitch attitude that will transfer the weight of the airplane from the wheels to the wings as rapidly as possible.
5. Lifts off at the lowest possible airspeed and remains in ground effect while accelerating to V_X or V_Y, as appropriate.
6. Establishes a pitch attitude for V_X or V_Y, as appropriate, and maintains selected airspeed ±5 knots, during the climb.
7. Retracts the landing gear, if appropriate and flaps after clear of any obstacles or as recommended by the manufacturer.

8. Maintains takeoff power and V_X or V_Y ±5 knots to a safe maneuvering altitude.
9. Maintains directional control and proper wind-drift correction throughout the takeoff and climb.
10. Completes appropriate checklist.

D. TASK: SOFT-FIELD APPROACH AND LANDING (ASEL)

REFERENCES: FAA-H-8083-3; POH/AFM.

Objective. To determine that the applicant:

1. Exhibits knowledge of the elements related to a soft-field approach and landing.
2. Considers the wind conditions, landing surface, and obstructions, and selects the most suitable touchdown area.
3. Establishes the recommended approach and landing configuration and airspeed; adjusts pitch attitude and power as required.
4. Maintains a stabilized approach and recommended airspeed, or in its absence, not more than 1.3 VSO, ±5 knots, with wind gust factor applied.
5. Makes smooth, timely, and correct control application during the roundout and touchdown.
6. Touches down softly, with no drift, and with the airplane's longitudinal axis aligned with the runway/landing path.
7. Maintains crosswind correction and directional control throughout the approach and landing sequence.
8. Maintains proper position of the flight controls and sufficient speed to taxi on the soft surface.
9. Completes appropriate checklist.

E. TASK: SHORT-FIELD TAKEOFF (CONFINED AREA—ASES) AND MAXIMUM PREFORMANCE CLIMB (ASEL and ASES)

REFERENCES: FAA-H-8083-3; POH/AFM.

Objective. To determine that the applicant:

1. Exhibits knowledge of the elements related to a short-field (confined area ASES) takeoff and maximum performance climb.
2. Positions the flight controls for the existing wind conditions, sets flaps as recommended.
3. Clears the area; taxies into takeoff position utilizing maximum available takeoff area and aligns the airplane on the runway center/takeoff path.

4. Selects an appropriate take-off path for the existing conditions (ASES).
5. Applies brakes (if appropriate) while advancing the throttle smoothly to takeoff power.
6. Establishes and maintains the most efficient planing/lift off attitude and corrects for porpoising and skipping (ASES).
7. Lifts off at the recommended airspeed, and accelerates to recommended obstacle clearance airspeed, or V_X.
8. Establishes a pitch attitude that will maintain the recommended obstacle clearance airspeed, or V_X+5/-0 knots, until the, obstacle is cleared, or until the airplane is 50 feet (20 meters) above the surface.
9. After clearing the obstacle, establishes the pitch attitude for V_Y, accelerates to V_Y, and maintains V_Y, ±5 knots, during the climb.
10. Retracts the landing gear, if appropriate and flaps after clear of any obstacles or as recommended by manufacturer.
11. Maintains takeoff power and V_Y ±5 knots to a safe maneuvering altitude.
12. Maintains directional control and proper wind-drift correction throughout the takeoff and climb.
13. Completes appropriate checklist.

F. TASK: SHORT-FIELD APPROACH (CONFINED AREA—ASES) AND LANDING (ASEL AND ASES)

REFERENCES: FAA-H-8083-3; POH/AFM.

Objective. To determine that the applicant:

1. Exhibits knowledge of the elements related to a short-field (confined area ASES) approach and landing.
2. Adequately surveys the intended landing area (ASES).
3. Considers the wind conditions, landing surface, obstructions, and selects the most suitable touchdown point.
4. Establishes the recommended approach and landing configuration and airspeed; adjusts pitch attitude and power.
5. Maintains a stabilized approach and recommended approach airspeed, or in its absence, not more than 1.3 V_{SO}, ±5 knots, with wind gust factor applied.
6. Makes smooth, timely, and correct control application during the roundout and touchdown.
7. Selects the proper landing path, contacts the water at the minimum safe airspeed with the proper pitch attitude for the surface conditions (ASES).
8. Touches down smoothly at minimum control airspeed (ASEL).

9. Touches down at or within 100 feet (30 meters) beyond a specified point, with no side drift, minimum float and with the airplane's longitudinal axis aligned with and over the runway center/landing path.
10. Maintains crosswind correction and directional control throughout the approach and landing sequence.
11. Applies brakes (ASEL) or elevator control (ASES), as necessary, to stop in the shortest distance consistent with safety.
12. Completes appropriate checklist.

G. TASK: GLASSY WATER TAKEOFF AND CLIMB (ASES)

NOTE: If a glassy water condition does not exist, the applicant shall be evaluated by simulating the TASK.

REFERENCES: FAA-H-8083-3; POH/AFM.

Objective. To determine that the applicant:

1. Exhibits knowledge of the elements related to glassy water takeoff and climb.
2. Positions the flight controls and flaps for the existing conditions.
3. Clears the area; selects an appropriate takeoff path considering surface hazards and/or vessels and surface conditions.
4. Retracts the water rudders as appropriate; advances the throttle smoothly to takeoff power.
5. Establishes and maintains an appropriate planing attitude, directional control, and corrects for porpoising, skipping, and increases in water drag.
6. Utilizes appropriate techniques to lift seaplane from the water considering surface conditions.
7. Establishes proper attitude/airspeed, and accelerates to V_Y, ± 5 knots during the climb.
8. Retracts the landing gear, if appropriate, and flaps after a positive rate of climb is established.
9. Maintains takeoff power and V_Y ±5 knots to a safe maneuvering altitude.
10. Maintains directional control and proper wind-drift correction throughout takeoff and climb.
11. Completes the appropriate checklist.

H. TASK: GLASSY WATER APPROACH AND LANDING (ASES)

NOTE: If a glassy water condition does not exist, the applicant shall be evaluated by simulating the TASK.

REFERENCES: FAA-H-8083-3; POH/AFM.

Objective. To determine that the applicant:

1. Exhibits knowledge of the elements related to glassy water approach and landing.
2. Adequately surveys the intended landing area.
3. Considers the wind conditions, water depth, hazards, surrounding terrain, and other watercraft.
4. Selects the most suitable approach path, and touchdown area.
5. Establishes the recommended approach and landing configuration and airspeed, and adjusts pitch attitude and power as required.
6. Maintains a stabilized approach and the recommended approach airspeed, $\pm$ 5 knots and maintains a touchdown pitch attitude and descent rate from the last altitude reference until touchdown.
7. Makes smooth, timely, and correct power and control adjustments to maintain proper pitch attitude and rate of descent to touchdown.
8. Contacts the water in the proper pitch attitude, and slows to idle taxi speed.
9. Maintains crosswind correction and directional control throughout the approach and landing sequence.
10. Completes the appropriate checklist.

I. TASK: ROUGH WATER TAKEOFF AND CLIMB (ASES)

NOTE: If a rough water condition does not exist, the applicant shall be evaluated by simulating the TASK.

REFERENCES: FAA-H-8083-3; POH/AFM.

Objective. To determine that the applicant:

1. Exhibits knowledge of the elements related to rough water takeoff and climb.
2. Positions the flight controls and flaps for the existing conditions.
3. Clears the area; selects an appropriate takeoff path considering wind, swells surface hazards and/or vessels.
4. Retracts the water rudders as appropriate; advances the throttle smoothly to takeoff power.

5. Establishes and maintains an appropriate planing attitude, directional control, and corrects for porpoising, skipping, or excessive bouncing.
6. Lifts off at minimum airspeed and accelerates to V_Y, ± 5 knots before leaving ground effect.
7. Retracts the landing gear, if appropriate, and flaps after a positive rate of climb is established.
8. Maintains takeoff power and V_Y ±5 knots to a safe maneuvering altitude.
9. Maintains directional control and proper wind-drift correction throughout takeoff and climb.
10. Completes the appropriate checklist.

J. TASK: ROUGH WATER APPROACH AND LANDING (ASES)

NOTE: If a rough water condition does not exist, the applicant shall be evaluated by simulating the TASK.

REFERENCES: FAA-H-8083-3; POH/AFM.

Objective. To determine that the applicant:

1. Exhibits knowledge of the elements related to rough water approach and landing.
2. Adequately surveys the intended landing area.
3. Considers the wind conditions, water, depth, hazards, surrounding terrain, and other watercraft.
4. Selects the most suitable approach path, and touchdown area.
5. Establishes the recommended approach and landing configuration and airspeed, and adjusts pitch attitude and power as required.
6. Maintains a stabilized approach and the recommended approach airspeed, or in its absence not more than 1.3 V_{SO} ± 5 knots with wind gust factor applied.
7. Makes smooth, timely, and correct power and control application during the roundout and touch down.
8. Contacts the water in the proper pitch attitude, and at the proper airspeed, considering the type of rough water.
9. Maintains crosswind correction and directional control throughout the approach and landing sequence.
10. Completes the appropriate checklist.

K. TASK: POWER-OFF 180 ACCURACY APPROACH AND LANDING (ASEL and ASES)

REFERENCE: FAA-H-8083-3.

Objective. To determine that the applicant:

1. Exhibits knowledge of the elements related to a power-off 180° accuracy approach and landing.
2. Considers the wind conditions, landing surface, obstructions, and selects an appropriate touchdown point.
3. Positions airplane on downwind leg, parallel to landing runway, and not more than 1000 feet AGL.
4. Abeam the specified touchdown point, closes throttle and establishes appropriate glide speed.
5. Completes final airplane configuration.
6. Touches down in a normal landing attitude, at or within 200 feet (60 meters) beyond the specified touchdown point.
7. Completes the appropriate checklist.

L. TASK: GO-AROUND/REJECTED LANDING (ASEL and ASES)

REFERENCES: FAA-H-8083-3; POH/AFM.

Objective. To determine that the applicant:

1. Exhibits knowledge of the elements related to a goaround/rejected landing.
2. Makes a timely decision to discontinue the approach to landing.
3. Applies takeoff power immediately and transitions to climb pitch attitude for V_Y, and maintains V_Y ±5 knots.
4. Retracts flaps as appropriate.
5. Retracts the landing gear if appropriate after a positive rate of climb is established.
6. Maneuvers to the side of runway/landing area to clear and avoid conflicting traffic.
7. Maintains takeoff power and V_Y ±5 knots to a safe maneuvering altitude.
8. Maintains directional control and proper wind-drift correction throughout the climb.
9. Completes the appropriate checklist.

V. AREA OF OPERATION: PERFORMANCE MANEUVERS

NOTE: The examiner shall at least select either TASK A or B, and either C or D.

A. TASK: STEEP TURNS (ASEL and ASES)

REFERENCES: FAA-H-8083-3; POH/AFM.

Objective. To determine that the applicant:

1. Exhibits knowledge of the elements related to steep turns.
2. Establishes the manufacturer's recommended airspeed or if one is not stated, a safe airspeed not to exceed VA.
3. Rolls into a coordinated 360° steep turn with at least a 50° bank, followed by a 360° steep turn in the opposite direction.
4. Divides attention between airplane control and orientation.
5. Maintains the entry altitude, ±100 feet (30 meters), airspeed, ±10 knots, bank, ±5°; and rolls out on the entry heading, ±10°.

B. TASK: STEEP SPIRAL (ASEL and ASES)

REFERENCE: FAA-H-8083-3.

Objective. To determine that the applicant:

1. Exhibits knowledge of the elements related to a steep spiral.
2. Selects an altitude sufficient to continue through a series of at least three 360° turns.
3. Selects a suitable ground reference point.
4. Applies wind-drift correction to track a constant radius circle around selected reference point with bank not to exceed 60° at steepest point in turn.
5. Divides attention between airplane control and ground track, while maintaining coordinated flight.
6. Maintains the specified airspeed, ±10 knots, rolls out toward object or specified heading, ±10°.

C. TASK: CHANDELLES (ASEL and ASES)

REFERENCE: FAA-H-8083-3.

Objective. To determine that the applicant:

1. Exhibits knowledge of the elements related to chandelles.
2. Selects an altitude that will allow the maneuver to be performed no lower than 1,500 feet AGL (460 meters).
3. Establishes the recommended entry configuration, power, and airspeed.
4. Establishes the angle of bank at approximately 30°.
5. Simultaneously applies power and pitch to maintain a smooth, coordinated climbing turn to the 90° point, with a constant bank.
6. Begins a coordinated constant rate rollout from the 90° point to the 180° point maintaining power and a constant pitch attitude.
7. Completes rollout at the 180° point, ±10° just above a stall airspeed, and maintaining that airspeed momentarily avoiding a stall.
8. Resumes straight and level flight with minimum loss of altitude.

D. TASK: LAZY EIGHTS (ASEL and ASES)

REFERENCE: FAA-H-8083-3.

Objective. To determine that the applicant:

1. Exhibits knowledge of the elements related to lazy eights.
2. Selects an altitude that will allow the task to be performed no lower than 1,500 feet AGL (460 meters).
3. Establishes the recommended entry configuration, power, and airspeed.
4. Maintains coordinated flight throughout the maneuver.
5. Achieves the following throughout the maneuver—

   a. approximately 30° bank at the steepest point.

   b. constant change of pitch and roll rate.

   c. altitude tolerance at 180° points, ±100 feet (30 meters) from entry altitude.

   d. airspeed tolerance at the 180° point plus ±10° knots from entry airspeed.

   e. heading tolerance at the 180° point ±10°.

6. Continues the maneuver through the number of symmetrical loops specified and resumes straight and level flight.

VI. AREA OF OPERATION: GROUND REFERENCE MANEUVER

TASK: EIGHTS ON PYLONS (ASEL and ASES)

REFERENCE: FAA-H-8083-3.

Objective. To determine that the applicant:

1. Exhibits knowledge of the elements related to eights on pylons.
2. Determines the approximate pivotal altitude.
3. Selects suitable pylons, that will permit straight and level flight, between the pylons.
4. Enters the maneuver at the appropriate altitude and airspeed and at a bank angle of approximately 30° to 40° at the steepest point.
5. Applies the necessary corrections so that the line-of-sight reference line remains on the pylon.
6. Divides attention between accurate coordinated airplane control and outside visual references.
7. Holds pylon using appropriate pivotal altitude avoiding slips and skids.

VII. AREA OF OPERATION: NAVIGATION

A. TASK: PILOTAGE AND DEAD RECKONING (ASEL and ASES)

REFERENCE: AC 61-23/FAA-H-8083-25.

Objective. To determine that the applicant:

1. Exhibits knowledge of the elements related to pilotage and dead reckoning.
2. Follows the preplanned course by reference to landmarks.
3. Identifies landmarks by relating surface features to chart symbols.
4. Navigates by means of precomputed headings, groundspeed, and elapsed time.
5. Corrects for and records differences between preflight groundspeed and heading calculations and those determined en route.
6. Verifies the airplane's position within two (2) nautical miles of flight planned route.
7. Arrives at the en route checkpoints within three (3) minutes of the initial or revised ETA and provides a destination estimate.
8. Maintains appropriate altitude, ±100 feet (30 meters), and headings, ±10°.

B. TASK: NAVIGATION SYSTEMS AND RADAR SERVICES (ASEL and ASES)

REFERENCES: FAA-H-8083-3, AC 61-23/FAA-H-8083-25; Navigation Equipment Operation Manuals, AIM.

Objective. To determine that the applicant:

1. Exhibits knowledge of the elements related to navigation systems and radar services.
2. Demonstrates the ability to use an airborne electronic navigation system.
3. Locates the airplane's position using the navigation system.
4. Intercepts and tracks a given course, radial, or bearing as appropriate.
5. Recognizes and describes the indication of station passage if appropriate.
6. Recognizes signal loss and takes appropriate action.
7. Uses proper communication procedures when utilizing radar services.
8. Maintains the appropriate altitude, ±100 feet (30 meters) and heading, ±10°.

C. TASK: DIVERSION (ASEL and ASES)

REFERENCES: FAA-H-8083-25; AIM

Objective. To determine that the applicant:

1. Exhibits knowledge of the elements related to diversion.
2. Selects an appropriate alternate airport and route.
3. Makes an accurate estimate of heading, groundspeed, arrival time, and fuel consumption to the alternate airport.
4. Maintains the appropriate altitude, ±100 feet (30 meters), and heading, ±10°.

D. TASK: LOST PROCEDURES (ASEL and ASES)

REFERENCES: FAA-H-8083-25; AIM.

Objective. To determine that the applicant:

1. Exhibits knowledge of the elements related to lost procedures.
2. Selects an appropriate course of action.
3. Maintains an appropriate heading and climbs, if necessary.
4. Identifies prominent landmarks.
5. Uses navigation systems/facilities and/or contacts an ATC facility for assistance as appropriate.

VIII. AREA OF OPERATION: SLOW FLIGHT AND STALLS

A. TASK: MANEUVERING DURING SLOW FLIGHT (ASEL and ASES)

REFERENCES: FAA-H-8083-3; POH/AFM.

Objective. To determine that the applicant:

1. Exhibits knowledge of the elements related to maneuvering during slow flight.
2. Selects an entry altitude that will allow the task to be completed no lower than 1,500 feet (460 meters) AGL.
3. Establishes and maintains an airspeed at which any further increase in angle of attack, increase in load factor, or reduction in power, would result in an immediate stall.
4. Accomplishes coordinated straight-and-level flight, turns, climbs, and descents with landing gear and flap configurations specified by the examiner.
5. Divides attention between airplane control and orientation.
6. Maintains the specified altitude, ±50 feet (15 meters); specified heading, ±10°; airspeed +5/−0 knots, and specified angle of bank, ±5°.

B. TASK: POWER-OFF STALLS (ASEL and ASES)

REFERENCES: FAA-H-8083-3, AC 61-67; POH/AFM.

Objective. To determine that the applicant:

1. Exhibits knowledge of the elements related to power-off stalls.
2. Selects an entry altitude that allows the task to be completed no lower than 1,500 feet (460 meters) AGL.
3. Establishes a stabilized descent in the approach or landing configuration, as specified by the examiner.
4. Transitions smoothly from the approach or landing attitude to a pitch attitude that will induce a stall.
5. Maintains a specified heading, ±10° in straight flight; maintains a specified angle of bank, not to exceed 20°, ±5°, in turning flight while inducing the stall.
6. Recognizes and recovers promptly as the stall occurs by simultaneously reducing the angle of attack, increasing power to maximum allowable, and leveling the wings to return to a straightand-level flight attitude with a minimum loss of altitude appropriate for the airplane.

7. Retracts the flaps to the recommended setting, retracts the landing gear if retractable after a positive rate of climb is established.
8. Accelerates to V_X or V_Y speed before the final flap retraction; returns to the altitude, heading, and airspeed specified by the examiner.

C. TASK: POWER-ON STALLS (ASEL and ASES)

REFERENCES: FAA-H-8083-3, AC 61-67; POH/AFM.

NOTE: In some high performance airplanes, the power setting may have to be reduced below the practical test standards guideline power setting to prevent excessively high pitch attitudes (greater than 30° nose up).

Objective. To determine that the applicant:

1. Exhibits knowledge of the elements related to power-on stalls.
2. Selects an entry altitude that allows the task to be completed no lower than 1,500 feet (460 meters) AGL.
3. Establishes the takeoff or departure configuration. Sets power to no less than 65 percent available power.
4. Transitions smoothly from the takeoff or departure attitude to a pitch attitude that will induce a stall.
5. Maintains a specified heading ±5°, in straight flight; maintains a specified angle of bank, not to exceed a 20°, ±10°, in turning flight, while inducing the stall.
6. Recognizes and recovers promptly as the stall occurs by simultaneously reducing the angle of attack, increasing power to maximum allowable, and leveling the wings to return to a straightand-level flight attitude, with a minimum loss of altitude appropriate for the airplane.
7. Retracts flaps to the recommended setting, retracts the landing gear if retractable, after a positive rate of climb is established.
8. Accelerates to V_X or V_Y speed before the final flap retraction; returns to the altitude, heading, and airspeed specified by the examiner.

D. TASK: SPIN AWARENESS (ASEL and ASES)

REFERENCES: FAA-H-8083-3, AC 61-67; POH/AFM

Objective. To determine that the applicant exhibits knowledge of the elements related to spin awareness by explaining:

1. Aerodynamic factors related to spins.
2. Flight situations where unintentional spins may occur.
3. Procedures for recovery from unintentional spins.

IX. AREA OF OPERATION: EMERGENCY OPERATIONS

A. TASK: EMERGENCY APPROACH AND LANDING (SIMULATED) (ASEL and ASES)

REFERENCES: FAA-H-8083-3; POH/AFM.

Objective. To determine that the applicant:

1. Exhibits knowledge of the elements related to emergency approach and landing procedures.
2. Analyzes the situation and selects an appropriate course of action.
3. Establishes and maintains the recommended best glide airspeed, ±10 knots.
4. Selects a suitable landing area.
5. Plans and follows a flight pattern to the selected landing area considering altitude, wind, terrain, and obstructions.
6. Prepares for landing, or go-around, as specified by the examiner.
7. Follows the appropriate checklist.

B. TASK: SYSTEMS AND EQUIPMENT MALFUNCTIONS (ASEL and ASES)

REFERENCES: FAA-H-8083-3; POH/AFM.

Objective. To determine that the applicant:

1. 1. Exhibits knowledge of the elements related to systems and equipment malfunctions appropriate to the airplane provided for the practical test.
2. 2. Analyzes the situation and takes appropriate action for simulated emergencies appropriate to the airplane provided for the practical test for at least five (5) of the following:

   a. partial or complete power loss.

   b. engine roughness or overheat.

   c. carburetor or induction icing.

   d. loss of oil pressure.

   e. fuel starvation.

   f. electrical malfunction.

   g. vacuum/pressure, and associated flight instruments malfunction.

   h. pitot/static.

i. landing gear or flap malfunction.

j. inoperative trim.

k. inadvertant door or window opening.

l. structural icing.

m. smoke/fire/engine compartment fire.

n. any other emergency appropriate to the airplane.

3. Follows the appropriate checklist or procedure.

C. TASK: EMERGENCY EQUIPMENT AND SURVIVAL GEAR (ASEL and ASES)

REFERENCES: FAA-H-8083-3; POH/AFM.

Objective. To determine that the applicant:

Exhibits knowledge of the elements related to emergency equipment and survival gear appropriate to the airplane and environment encountered during flight. Identifies appropriate equipment that should be aboard the airplane.

X. AREA OF OPERATION: HIGH ALTITUDE OPERATIONS

A. TASK: SUPPLEMENTAL OXYGEN (ASEL and ASES)

REFERENCES: 14 CFR part 91; FAA-H-8083-3, AC 61-107; AIM, POH/AFM.

Objective. To determine that the applicant exhibits knowledge of the elements related to supplemental oxygen by explaining:

Supplemental oxygen requirements for flight crew and passengers when operating non-pressurized airplanes.

1. Identification and differences between "aviators" breathing oxygen" and other types.
2. perational characteristics of continuous flow, demand, and pressure-demand oxygen systems.

B. TASK: PRESSURIZATION (ASEL and ASES)

REFERENCES: FAA-H-8083-3, AC 61-107; AIM, POH/AFM.

Objective. To determine that the applicant:

1. Exhibits knowledge of the elements related to pressurization by explaining—
   a. fundamental concept of cabin pressurization.
   b. supplemental oxygen requirements when operating airplanes with pressurized cabins.
   c. physiological hazards associated with high altitude flight and decompression.

NOTE: Element 2 applies only if the airplane provided for the practical test is equipped for pressurized flight operations.

2. Operates the pressurization system properly, and reacts appropriately to simulated pressurization malfunctions.

XI. AREA OF OPERATION: POSTFLIGHT PROCEDURES

NOTE: The examiner shall select TASK A and for AMES applicants at least one other TASK.

A. TASK: AFTER LANDING, PARKING, AND SECURING (ASEL and ASES)

REFERENCES: FAA-H-8083-3; POH/AFM.

Objective. To determine that the applicant:

1. Exhibits knowledge of the elements related to after landing, parking and securing procedures.
2. Maintains directional control after touchdown while decelerating to an appropriate speed.
3. Observes runway hold lines and other surface control markings and lighting.
4. Parks in an appropriate area, considering the safety of nearby persons and property.
5. Follows the appropriate procedure for engine shutdown.
6. Completes the appropriate checklist.
7. Conducts an appropriate postflight inspection and secures the aircraft.

B. TASK: ANCHORING (ASES)

REFERENCES: FAA-H-8083-3; POH/AFM.

Objective. To determine that the applicant:

1. Exhibits knowledge of the elements related to anchoring.
2. Selects a suitable area for anchoring, considering seaplane movement, water depth, tide, wind, and weather changes.
3. Uses an adequate number of anchors and lines of sufficient strength and length to ensure the seaplane's security.

C. TASK: DOCKING AND MOORING (ASES)

REFERENCES: FAA-H-8083-3; POH/AFM.

Objective. To determine that the applicant:

1. Exhibits knowledge of the elements related to docking and mooring.
2. Approaches the dock or mooring buoy in the proper direction considering speed, hazards, wind, and water current.
3. Ensures seaplane security.

D. TASK: RAMPING/BEACHING (AMES)

REFERENCES: FAA-H-8083-3; POH/AFM.

Objective. To determine that the applicant:

1. Exhibits knowledge of the elements related to ramping/beaching.
2. Approaches the ramp/beach considering persons and property, in the proper attitude and direction, at a safe speed, considering water depth, tide, current, and wind.
3. Ramps/beaches and secures the seaplane in a manner that will protect it from the harmful effect of wind, waves, and changes in water level.

APPENDIX 1

TASK VS. SIMULATION DEVICE CREDIT

Single-Engine Land (SEL)

APPENDIX 1

AIRPLANE SINGLE-ENGINE LAND TASK VS. SIMULATION DEVICE CREDIT

Examiners conducting the Commercial Pilot–Airplane Practical Tests with flight simulation devices should consult appropriate documentation to ensure that the device has been approved for training, testing, or checking. The documentation for each device should reflect that the following activities have occurred:

1. The device must be evaluated, determined to meet the appropriate standards, and assigned the appropriate qualification level by the National Simulator Program Manager. The device must continue to meet qualification standards through continuing evaluations as outlined in the appropriate advisory circular (AC). For airplane flight training devices (FTDs), AC 120-45 (as amended), Airplane Flight Training Device Qualifications, will be used. For simulators, AC 120-40 (as amended), Airplane Simulator Qualification, will be used.

2. The FAA must approve the device for training, testing, and checking the specific flight TASKs listed in this appendix.

3. The device must continue to support the level of student or applicant performance required by this PTS.

NOTE: Users of the following chart are cautioned that use of the chart alone is incomplete. The description and objective of each TASK as listed in the body of the PTS, including all notes, must also be incorporated for accurate simulation device use.

USE OF CHART

X Creditable.

A Creditable if appropriate systems are installed and operating.

* Asterisk items require use of FTD or simulator visual reference.

NOTES: 1. Use of Level 2 or Level 3 FTDs authorized only for those airplanes not requiring a type rating.

2. For practical tests, not more than 50 % of the maneuvers may be accomplished in an FTD or simulator UNLESS:

a. each maneuver has been satisfactorily accomplished for an instructor, in the appropriate airplane, not less than three (3) times,

OR

b. the applicant has logged not less than 500 hours of flight time as a pilot in airplanes.

3. Not all AREAS OF OPERATION (AOO) and TASKs required by this PTS are listed in the appendix. The remaining AOO and TASKs must be accomplished in an airplane.

4. Standards for and use of Level 1 FTD's have not been determined.

APPENDIX 1

FLIGHT TASK Areas of Operation:	AIRPLANE SINGLE-ENGINE LAND FLIGHT SIMULATION DEVICE LEVEL 1	2	3	4	5	6	7	A	B	C	D
II. Preflight Procedures											
A. Preflight Inspection (Cockpit Only)		A	X	A	A	X	X	X	X	X	X
B. Cockpit Management		A	X	A	A	X	X	X	X	X	X
C. Engine Starting		A	X	A	A	X	X	X	X	X	X
D. Taxiing										X	X
E. Before Takeoff Check		A	X	A	A	X	X	X	X	X	X
IV. Takeoffs, Landings, and Go-Arounds											
A. Normal and Crosswind Takeoff and Climb										X	X
B. Normal and Crosswind Approach and Landing										X	X
E. Short-Field Takeoff and Climb								X	X	X	X
F. Short-Field Approach and Landing										X	X
G. Go-Around *			X			X	X	X	X	X	X
V. Performance Maneuvers											
A. Steep Turns			X			X	X	X	X	X	X
VII. Navigation *											
B. Navigation Systems and ATC Radar Services		A			A	X	X	X	X	X	X
C. Diversion		A	X		A	X	X	X	X	X	X
D. Lost Procedures		A	X		A	X	X	X	X	X	X
VIII. Slow Flight and Stalls											
A. Maneuvering During Slow Flight			X			X	X	X	X	X	X
IX. Emergency Operations											
A. Emergency Descent			X			X	X	X	X	X	X
B. Emergency Approach and Landing										X	X
C. Systems and Equipment Malfunctions		A	X	A	A	X	X	X	X	X	X
X. High Altitude Operations											
B. Pressurization		A	X	A	A	X	X	X	X	X	X
XI. Postflight Procedures											
A. After Landing		A	X	A	A	X	X	X	X	X	X

SECTION 2

COMMERCIAL PILOT—AIRPLANE

Multiengine Land and Multiengine Sea

CONTENTS

Airplane Multiengine Land and Airplane Multiengine Sea

ADDITIONAL RATING TASK TABLE

Airplane Multiengine Land

Addition of an Airplane Multiengine Land Rating to an existing Commercial Pilot Certificate								
Required TASKs are indicated by either the TASK letter(s) that apply(s) or an indication that all or none of the TASKs must be tested based on the notes in each AREA OF OPERATION.								
COMMERCIAL PILOT RATING(S) HELD								
AREAS OF OPER-ATION	ASEL	ASES	AMES	RH	RG	Glider	Balloon	Airship
I	F,G,H	F,G,H	F,G	F,G,H	F,G,H	F,G,H	F,G,H	F,G,H
II	ALL	ALL	D	ALL	ALL	ALL	ALL	ALL
III	NONE	C	C	B,C	NONE	B,C	B,C	B,C
IV	A,B,C, D	A,B,C, D	A,B,C, D	A,B,C, D,I	A,B,C, D,I	A,B,C, D,I	A,B,C, D,I	A,B,C, D,I
V	ALL	ALL	NONE	ALL	ALL	ALL	ALL	ALL
VI	NONE	NONE	NONE	NONE	NONE	ALL	ALL	NONE
VII	ALL	ALL	NONE	ALL	ALL	ALL	ALL	ALL
VIII	ALL	ALL	B,D,E	ALL	ALL	ALL	ALL	ALL
IX	NONE	NONE	NONE	ALL	ALL	ALL	ALL	ALL
X	ALL	ALL	NONE	ALL	ALL	ALL	ALL	ALL
XI	NONE	A	A	A	A	A	A	A

ADDITIONAL RATING TASK TABLE

Airplane Multiengine Sea

Addition of an Airplane Multiengine Sea Rating to an existing Commercial Pilot Certificate								
Required TASKs are indicated by either the TASK letter(s) that apply(s) or an indication that all or none of the TASKs must be tested based on the notes in each AREA OF OPERATION.								
COMMERCIAL PILOT RATING(S) HELD								
AREAS OF OPER-ATION	**AMEL**	**ASEL**	**ASES**	**RH**	**RG**	**Glider**	**Balloon**	**Airship**
I	F,G,I,J	F,G,H, I,J	F,G,H	F,G,H, I,J	F,G,H, I,J	F,G,H, I,J	F,G,H, I,J	F,G,H, I,J
II	E	ALL	ALL	ALL	ALL	ALL	ALL	ALL
III	C	C	NONE	B,C	C	B,C	B,C	B,C
IV	A,B,C, D,E,F, G,H	A,B,C, D,E,F, G,H	A,B,C, D,E,F, G,H	ALL	ALL	ALL	ALL	ALL
V	NONE	ALL	ALL	ALL	ALL	ALL	ALL	ALL
VI	NONE	NONE	NONE	NONE	NONE	ALL	ALL	NONE
VII	NONE	ALL	ALL	ALL	ALL	ALL	ALL	ALL
VIII	B,D,E	ALL	ALL	ALL	ALL	ALL	ALL	ALL
IX	NONE	NONE	NONE	ALL	ALL	ALL	ALL	ALL
X	NONE	ALL	ALL	ALL	ALL	ALL	ALL	ALL
XI	B,C,D	B,C,D	NONE	ALL	B,C,D	ALL	ALL	ALL

APPLICANT'S PRACTICAL TEST CHECKLIST

APPOINTMENT WITH EXAMINER:

EXAMINER'S NAME____

LOCATION ____

DATE/TIME ____

ACCEPTABLE AIRCRAFT

□ Aircraft Documents:

Airworthiness Certificate
Registration Certificate
Operating Limitations

□ Aircraft Maintenance Records:

Logbook Record of Airworthiness Inspections and AD Compliance

□ Pilot's Operating Handbook, FAA-Approved

Airplane Flight Manual

PERSONAL EQUIPMENT

□ View-Limiting Device

□ Current Aeronautical Charts

□ Computer and Plotter

□ Flight Plan Form

□ Flight Logs

□ Current AIM, Airport Facility Directory, and Appropriate Publications

PERSONAL RECORDS

□ Identification—Photo/Signature ID

□ Pilot Certificate

□ Current and Appropriate Medical Certificate

- □ Completed FAA Form 8710-1, Airman Certificate and/or Rating Application with Instructor's Signature (If applicable)
- □ Computer Test Report
- □ Pilot Logbook with appropriate Instructor Endorsements
- □ FAA Form 8060-5, Notice of Disapproval (if applicable)
- □ Approved School Graduation Certificate (if applicable)
- □ Examiner's Fee (if applicable)

EXAMINER'S PRACTICAL TEST CHECKLIST

Airplane Single-Engine Land And Airplane Single-Engine Sea

APPLICANT'S NAME________________________________

LOCATION______________________________________

DATE/TIME_____________________________________

I. PREFLIGHT PREPARATION

□ A. Certificates and Documents (AMEL and AMES)

□ **B.** Airworthiness Requirements (AMEL and AMES)

□ **C.** Weather Information (AMEL and AMES)

□ **D.** Cross-Country Flight Planning (AMEL and AMES)

□ **E.** National Airspace System (AMEL and AMES)

□ **F.** Performance and Limitations (AMEL and AMES)

□ **G.** Operation of Systems (AMEL and AMES)

□ **H.** Water and Seaplane Characteristics (AMES)

□ **I.** Seaplane Bases, Maritime Rules, and Aids to Marine Navigation (AMES)

□ **J.** Aeromedical Factors (AMEL and AMES)

II. PREFLIGHT PROCEDURES

□ **A.** Preflight Inspection (AMEL and AMES)

□ **B.** Cockpit Management (AMEL and AMES)

□ **C.** Engine Starting (AMEL and AMES)

□ **D.** Taxiing (ASEL)

- ☐ **E.** Taxiing and Sailing (AMES)
- ☐ **F.** Before Takeoff Check (AMEL and AMES)

III. AIRPORT AND SEAPLANE BASE OPERATIONS

- ☐ **A.** Radio Communications and ATC Light Signals (AMEL and AMES)
- ☐ **B.** Traffic Patterns (AMEL and AMES)
- ☐ **C.** Airport/Seaplane Base, Runway, and Taxiway Signs, Markings, and Lighting (AMEL and AMES)

IV. TAKEOFFS, LANDINGS, AND GO-AROUNDS

- ☐ **A.** Normal and Crosswind Takeoff and Climb (AMEL and AMES)
- ☐ **B.** Normal and Crosswind Approach and Landing (AMEL and AMES)
- ☐ **C.** Short-Field (Confined Area—ASES) Takeoff and Maximum Performance Climb (AMEL and AMES)
- ☐ **D.** Short-Field Approach (Confined Area—ASES) and Landing (AMEL and AMES)
- ☐ **E.** Glassy Water Takeoff and Climb (AMES)
- ☐ **F.** Glassy Water Approach and Landing (AMES)
- ☐ **G.** Rough Water Takeoff and Climb (AMES)
- ☐ **H.** Rough Water Approach and Landing (AMES)
- ☐ **I.** Go-Around/Rejected Landing (AMEL and AMES)

V. PERFORMANCE MANEUVER

- ☐ Steep Turns (AMEL and AMES)

VI. NAVIGATION

- ☐ **A.** Pilotage and Dead Reckoning (AMEL and AMES)
- ☐ **B.** Navigation Systems and Radar Services (AMEL and AMES)
- ☐ **C.** Diversion (AMEL and AMES)
- ☐ **D.** Lost Procedures (AMEL and AMES)

VII. SLOW FLIGHT AND STALLS

- ☐ **A.** Maneuvering During Slow Flight (AMEL and AMES)
- ☐ **B.** Power-Off Stalls (AMEL and AMES)
- ☐ **C.** Power-On Stalls (AMEL and AMES)
- ☐ **D.** Spin Awareness (AMEL and AMES)

VIII. EMERGENCY OPERATIONS

- ☐ **A.**Emergency Descent (AMEL and AMES)
- ☐ **B.** Engine Failure During Takeoff Before V_{MC} (Simulated) (AMEL and AMES)
- ☐ **C.** Engine Failure After Lift-Off (Simulated) (AMEL and AMES)
- ☐ **D.** Approach and Landing with an Inoperative Engine (Simulated) (AMEL and AMES)
- ☐ **E.** Systems and Equipment Malfunctions (AMEL and AMES)
- ☐ **F.** Emergency Equipment and Survival Gear (AMEL and AMES)

IX. HIGH ALTITUDE OPERATIONS

- ☐ **A.** Supplemental Oxygen (AMEL and AMES)
- ☐ **B.** Pressurization (AMEL and AMES)

X. MULTIENGINE OPERATIONS

- ☐ **A.** Maneuvering with One Engine Inoperative (AMEL and AMES)
- ☐ **B.** Vmc Demonstration (AMEL and AMES)
- ☐ **C.** Engine Failure During Flight (by Reference to Instruments) (AMEL and AMES)
- ☐ **D.** Instrument Approach—One Engine Inoperative (by Reference to Instruments) (AMEL and AMES)

XI. POSTFLIGHT PROCEDURES

- ☐ **A.** After Landing, Parking, and Securing (AMEL and AMES)

- ☐ **B.** Anchoring (AMES)
- ☐ **C.** Docking and Mooring (AMES)
- ☐ **D.** Ramping/Beaching (AMES)

I. AREA OF OPERATION: PREFLIGHT PREPARATION

NOTE: The examiner shall develop a scenario based on real time weather to evaluate TASKs C and D.

A. TASK: CERTIFICATES AND DOCUMENTS (AMEL and AMES)

REFERENCES: 14 CFR parts 43, 61, 91; FAA-H-8083-3, AC 6123/ FAA-H-8083-25; POH/AFM.

Objective. To determine that the applicant exhibits knowledge of the elements related to certificates and documents by:

1. Explaining—
   a. commercial pilot certificate privileges limitations and recent flight experience requirements.
   b. medical certificate class and duration.
   c. pilot logbook or flight records.
2. Locating and explaining—
   a. airworthiness and registration certificates.
   b. operating limitations, placards, instrument markings, and POH/AFM.
   c. weight and balance data and equipment list.

B. TASK: AIRWORTHINESS REQUIREMENTS (AMEL and AMES)

REFERENCES: 14 CFR part 91; AC 61-23/FAA-H-8083-25

Objective. To determine that the applicant exhibits knowledge of the elements related to airworthiness requirements by:

1. Explaining—
   a. required instruments and equipment for day/night VFR.
   b. procedures and limitations for determining airworthiness of the airplane with inoperative instruments and equipment with and without an MEL.
   c. requirements and procedures for obtaining a special flight permit.
2. Locating and explaining—
   a. airworthiness directives.
   b. compliance records.
   c. maintenance/inspection requirements.
   d. appropriate record keeping.

C. TASK: WEATHER INFORMATION (AMEL and AMES)

REFERENCES: 14 CFR part 91; AC 00-6, AC 00-45, AC 61-23/FAA-H8083-25, AC 61-84; AIM.

Objective. To determine that the applicant:

1. Exhibits knowledge of the elements related to weather information by analyzing weather reports, charts, and forecasts from various sources with emphasis on—
   a. METAR, TAF, and FA.
   b. surface analysis chart.
   c. radar summary chart.
   d. winds and temperature aloft chart.
   e. significant weather prognostic charts.
   f. convective outlook chart.
   g. AWOS, ASOS, and ATIS reports.
2. Makes a competent "go/no-go" decision based on available weather information.

D. TASK: CROSS-COUNTRY FLIGHT PLANNING (AMEL and AMES)

REFERENCES: 14 CFR part 91; AC 61-23/FAA-H-8083-25, AC 61-84; Navigation Charts; A/FD; AIM.

Objective. To determine that the applicant:

1. Exhibits knowledge of the elements related to cross-country flight planning by presenting and explaining a pre-planned VFR cross-country flight, as previously assigned by the examiner. On the day of the practical test, the final flight plan shall be to the first fuel stop, based on maximum allowable passengers, baggage and/or cargo loads using real time weather.
2. Uses appropriate and current aeronautical charts.
3. Properly identifies airspace, obstructions, and terrain features.
4. Selects easily identifiable en route checkpoints.
5. Selects most favorable altitudes considering weather conditions and equipment capabilities.
6. Computes headings, flight time, and fuel requirements.
7. Selects appropriate navigation system/facilities and communication frequencies.
8. Applies pertinent information from NOTAMs, A/FD, and other flight publications.
9. Completes a navigation log and simulates filing a VFR flight plan.

E. TASK: NATIONAL AIRSPACE SYSTEM (AMEL and AMES)

REFERENCES: 14 CFR parts 71, 91; Navigation Charts; AIM.

Objective. To determine that the applicant exhibits knowledge of the elements related to the National Airspace System by explaining:

1. Basic VFR weather minimums—for all classes of airspace.
2. Airspace classes—their operating rules, pilot certification, and airplane equipment requirements for the following—
   a. Class A.
   b. Class B.
   c. Class C.
   d. Class D.
   e. Class E.
   f. Class G.
3. Special use and other airspace areas.

F. TASK: PERFORMANCE AND LIMITATIONS (AMEL and AMES)

REFERENCES: AC 61-23/FAA-H-8083-25, FAA-H-8083-1, AC 61-84; POH/AFM.

Objective. To determine that the applicant:

1. Exhibits knowledge of the elements related to performance and limitations by explaining the use of charts, tables, and data to determine performance and the adverse effects of exceeding limitations.
2. Computes weight and balance. Determines if the computed weight and center of gravity is within the airplane's operating limitations and if the weight and center of gravity will remain within limits during all phases of flight.
3. Demonstrates use of the appropriate performance charts, tables, and data.
4. Describes the effects of atmospheric conditions on the airplane's performance.

G. TASK: OPERATION OF SYSTEMS (AMEL and AMES)

REFERENCES: AC 61-23/FAA-H-8083-25; POH/AFM.

Objective. To determine that the applicant exhibits knowledge of the elements related to the operation of systems on the airplane provided for the practical test, by explaining at least five (5) of the following systems:

1. Primary flight controls and trim.
2. Flaps, leading edge devices, and spoilers.

3. Water rudders (ASES).
4. Powerplant and propeller.
5. Landing gear.
6. Fuel, oil, and hydraulic.
7. Electrical.
8. Avionics.
9. Pitot-static, vacuum/pressure and associated flight instruments.
10. Environmental.
11. Deicing and anti-icing.

H. TASK: PRINCIPLES OF FLIGHT—ENGINE INOPERATIVE (AMEL and AMES)

REFERENCES: FAA-H-8083-3, AC 61-23/FAA-H-8083-25; POH/AFM.

Objective. To determine that the applicant exhibits knowledge of the elements related to engine inoperative principles of flight by explaining the:

1. meaning of the term "critical engine."
2. effects of density altitude on the V_{MC} demonstration.
3. effects of airplane weight and center of gravity on control.
4. effects of angle of bank on VMC.
5. relationship of VMC to stall speed.
6. reasons for loss of directional control.
7. indications of loss of directional control.
8. importance of maintaining the proper pitch and bank attitude, and the proper coordination of controls.
9. loss of directional control recovery procedure.
10. engine failure during takeoff including planning, decisions, and single-engine operations.

I. TASK: WATER AND SEAPLANE CHARACTERISTICS (AMES)

REFERENCE: FAA-H-8083-3.

Objective. To determine that the applicant exhibits knowledge of the elements related to water and seaplane characteristics by explaining:

1. The characteristics of a water surface as affected by features, such as—
    a. size and location.
    b. protected and unprotected areas.

c. surface wind.
d. direction and strength of water current.
e. floating and partially submerged debris.
f. sandbars, islands, and shoals.
g. vessel traffic and wakes.
h. other features peculiar to the area.

2. Float and hull construction, and their effect on seaplane performance.
3. Causes of porpoising and skipping, and the pilot action required to prevent or correct these occurrences.

J. TASK: SEAPLANE BASES, MARITIME RULES, AND AIDS TO MARINE NAVIGATION (AMES)

REFERENCES: FAA-H-8083-3; AIM.

Objective. To determine that the applicant exhibits knowledge of the elements related to seaplane bases, maritime rules, and aids to marine navigation by explaining:

1. How to locate and identify seaplane bases on charts or in directories.
2. Operating restrictions at various bases.
3. Right-of-way, steering, and sailing rules pertinent to seaplane operation.
4. Marine navigation aids, such as buoys, beacons, lights, and sound signals.

K. TASK: AEROMEDICAL FACTORS (AMEL and AMES)

REFERENCES: AC 61-23/FAA/H-8083-25; AIM.

Objective. To determine that the applicant exhibits knowledge of the elements related to aeromedical factors by explaining:

1. The symptoms, causes, effects, and corrective actions of at least four (4) of the following—
    a. hypoxia.
    b. hyperventilation.
    c. middle ear and sinus problems.
    d. spatial disorientation.
    e. motion sickness.
    f. carbon monoxide poisoning.
    g. stress and fatigue.
    h. dehydration.
2. The effects of alcohol, drugs, and over-the-counter medications.
3. The effects of excess nitrogen during scuba dives upon a pilot or passenger in flight.

II. AREA OF OPERATION: PREFLIGHT PROCEDURES

A. TASK: PREFLIGHT INSPECTION (AMEL and AMES)

REFERENCES: FAA-H-8083-3; POH/AFM.

Objective. To determine that the applicant:

1. Exhibits knowledge of the elements related to preflight inspection. This shall include which items must be inspected, the reasons for checking each item, and how to detect possible defects.
2. Inspects the airplane with reference to an appropriate checklist.
3. Verifies that the airplane is in condition for safe flight.

B. TASK: COCKPIT MANAGEMENT (AMEL and AMES)

REFERENCES: FAA-H-8083-3; POH/AFM.

Objective. To determine that the applicant:

1. Exhibits knowledge of the elements related to cockpit management procedures.
2. Ensures all loose items in the cockpit and cabin are secured.
3. Organizes material and equipment in an efficient manner so they are readily available.
4. Briefs occupants on the use of safety belts, shoulder harnesses, doors, and emergency procedures.

C. TASK: ENGINE STARTING (AMEL and AMES)

REFERENCES: FAA-H-8083-3, AC 61-23/FAA-H-8083-25, AC 91-13, AC 91-55; POH/AFM.

Objective. To determine that the applicant:

1. Exhibits knowledge of the elements related to recommended engine starting procedures. This shall include the use of an external power source, and starting under various atmospheric conditions.
2. Positions the airplane properly considering structures, surface conditions, other aircraft, and the safety of nearby persons and property.
3. Utilizes the appropriate checklist for starting procedure.

D. TASK: TAXIING (AMEL)

REFERENCES: FAA-H-8083-3; POH/AFM.

Objective. To determine that the applicant:

1. Exhibits knowledge of the elements related to safe taxi procedures.
2. Performs a brake check immediately after the airplane begins moving.
3. Positions flight controls properly for the existing wind conditions.
4. Controls direction and speed without excessive use of brakes.
5. Complies with airport/taxiway markings, signals, ATC clearances, and instructions.
6. Taxies so as to avoid other aircraft and hazards.

E. TASK: TAXIING AND SAILING (AMES)

REFERENCES: FAA-H-8083-3; USCG NAVIGATION RULES, INTERNATIONAL–INLAND; POH/AFM.

Objective. To determine that the applicant:

1. Exhibits knowledge of the elements related to water taxi and sailing procedures.
2. Positions the flight controls properly for the existing wind conditions.
3. Plans and follows the most favorable course while taxi or sailing considering wind, water current, water conditions and maritime regulations.
4. Uses the appropriate idle, plow, or step taxi technique.
5. Uses flight controls, flaps, doors, water rudder, and power correctly so as to follow the desired course while sailing.
6. Prevents and corrects for porpoising and skipping.
7. Avoids other aircraft, vessels, and hazards.
8. Complies with seaplane base signs, signals, and clearances.

F. TASK: BEFORE TAKEOFF CHECK (AMEL and AMES)

REFERENCES: FAA-H-8083-3; POH/AFM.

Objective. To determine that the applicant:

1. Exhibits knowledge of the elements related to the before takeoff check. This shall include the reasons for checking each item and how to detect malfunctions.

2. Positions the airplane properly considering other aircraft/vessels, wind and surface conditions.
3. Divides attention inside and outside the cockpit.
4. Ensures the engine temperatures and pressure are suitable for run-up and takeoff.
5. Accomplishes the before takeoff checklist and ensures the airplane is in safe operating condition.
6. Reviews takeoff performance airspeeds, takeoff distances, departures and emergency procedures.
7. Avoids runway incursion and/or ensures no conflict with traffic prior to taxiing into takeoff position.

III. AREA OF OPERATION: AIRPORT AND SEAPLANE BASE OPERATIONS

A. TASK: RADIO COMMUNICATIONS AND ATC LIGHT SIGNALS (AMEL and AMES)

REFERENCES: 14 CFR part 91; AC 61-23/FAA-H-8083-25; AIM.

Objective. To determine that the applicant:

1. Exhibits knowledge of the elements related to radio communications and ATC light signals.
2. Selects appropriate frequencies.
3. Transmits using recommended phraseology.
4. Acknowledges radio communications and complies with instructions.

B. TASK: TRAFFIC PATTERNS (AMEL and AMES)

REFERENCES: FAA-H-8083-3, AC 61-23/FAA-H-8083-25, AC 90-66; AIM.

Objective. To determine that the applicant:

1. Exhibits knowledge of the elements related to traffic patterns. This shall include procedures at airports with and without operating control towers, prevention of runway incursions, collision avoidance, wake turbulence avoidance, and wind shear.
2. Complies with proper traffic pattern procedures.
3. Maintains proper spacing from other aircraft.
4. Corrects for wind-drift to maintain proper ground track.
5. Maintains orientation with runway/landing area in use.
6. Maintains traffic pattern altitude ±100 feet (30 meters), and appropriate airspeed ±10 knots.

C. TASK: AIRPORT/SEAPLANE BASE, RUNWAY, AND TAXIWAY SIGNS, MARKINGS, AND LIGHTING (AMEL and AMES)

REFERENCES: AC 61-23/FAA-H-8083-25; AIM.

Objective. To determine that the applicant:

1. Exhibits knowledge of the elements related to airport/seaplane base, runway, and taxiway operations with emphasis on runway incursion avoidance.
2. Properly identifies and interprets airport/seaplane base, runway, and taxiway signs, markings, and lighting.

IV. AREA OF OPERATION: TAKEOFFS, LANDINGS, AND GO-AROUNDS

A. TASK: NORMAL AND CROSSWIND TAKEOFF AND CLIMB (AMEL and AMES)

NOTE: If a crosswind condition does not exist, the applicant's knowledge of crosswind elements shall be evaluated through oral testing.

REFERENCES: FAA-H-8083-3; POH/AFM.

Objective. To determine that the applicant:

1. Exhibits knowledge of the elements related to normal and crosswind takeoff, climb operations, and rejected takeoff procedures.
2. Positions the flight controls for the existing wind conditions.
3. Clears the area, taxies onto the takeoff surface and aligns the airplane on the runway center/takeoff path.
4. Retracts the water rudders as appropriate, (AMES) advances the throttles smoothly to takeoff power.
5. Establishes and maintains the most efficient planing/lift off attitude and corrects for porpoising and skipping (AMES).
6. Lifts off at the recommended airspeed and accelerates to V_Y.
7. Establishes a pitch attitude that will maintain V_Y ±5 knots.
8. Retracts the landing gear, if appropriate, and flaps after a positive rate of climb is established.
9. Maintains takeoff power and V_Y ±5 knots to a safe maneuvering altitude.
10. Maintains directional control, proper wind-drift correction throughout the takeoff and climb.
11. Complies with noise abatement procedures.
12. Completes appropriate checklists.

B. TASK: NORMAL AND CROSSWIND APPROACH AND LANDING (AMEL and AMES)

NOTE: If a crosswind condition does not exist, the applicant's knowledge of the crosswind elements shall be evaluated through oral testing.

REFERENCES: FAA-H-8083-3; POH/AFM.

Objective. To determine that the applicant:

1. Exhibits knowledge of the elements related to normal and crosswind approach and landing.
2. Adequately surveys the intended landing area (AMES).
3. Considers the wind conditions, landing surface, obstructions, and selects a suitable touchdown point.
4. Establishes the recommended approach and landing configuration and airspeed and adjusts pitch attitude and power as required.
5. Maintains a stabilized approach and recommended airspeed, or in its absence, not more than 1.3 V_{SO}, ±5 knots, with wind gust factor applied.
6. Makes smooth, timely, and correct control application during the roundout and touchdown.
7. Contacts the water at the proper pitch attitude (AMES).
8. Touches down smoothly at approximate stalling speed (AMEL).
9. Touches down at or within 200 feet (60 meters) beyond a specified point, with no drift, and with the airplane's longitudinal axis aligned with and over the runway center/landing path.
10. Maintains crosswind correction and directional control throughout the approach and landing sequence.
11. Completes appropriate checklist.

C. TASK: SHORT-FIELD TAKEOFF (CONFINED AREA—AMEL) AND MAXIMUM PREFORMANCE CLIMB (AMEL and AMES)

REFERENCES: FAA-H-8083-3; POH/AFM.

Objective. To determine that the applicant:

1. Exhibits knowledge of the elements related to a short-field confined area (AMES) takeoff and maximum performance climb.
2. Positions the flight controls for the existing wind conditions, sets flaps as recommended.
3. Clears the area; taxies into takeoff position utilizing maximum available takeoff area and aligns the airplane on the runway center/takeoff path.
4. Selects an appropriate take-off path for the existing conditions (AMES).
5. Applies brakes (if appropriate) while advancing the throttles smoothly to takeoff power.
6. Establishes and maintains the most efficient planing/lift off attitude and corrects for porpoising and skipping (AMES).
7. Lifts off at the recommended airspeed, and accelerates to recommended obstacle clearance airspeed, or V_X.

8. Establishes a pitch attitude that will maintain the recommended obstacle clearance airspeed, or V_X, +5/-0 knots, until the obstacle is cleared, or until the airplane is 50 feet (20 meters) above the surface.
9. After clearing the obstacle, establishes the pitch attitude for V_Y, accelerates to V_Y, and maintains V_Y, ±5 knots, during the climb.
10. Retracts the landing gear, if appropriate, and flaps after clear of any obstacles or as recommended by manufacturer.
11. Maintains takeoff power and V_Y ±5 knots to a safe maneuvering altitude.
12. Maintains directional control and proper wind-drift correction throughout the takeoff and climb.
13. Completes appropriate checklist.

D. TASK: SHORT-FIELD (CONFINED AREA—AMES) APPROACH AND LANDING (AMEL and AMES)

REFERENCES: FAA-H-8083-3; POH/AFM.

Objective. To determine that the applicant:

1. Exhibits knowledge of the elements related to a short-field (confined area AMES) approach and landing.
2. Adequately surveys the intended landing area (AMES).
3. Considers the wind conditions, landing surface, obstructions, and selects the most suitable touchdown point.
4. Establishes the recommended approach and landing configuration and airspeed; adjusts pitch attitude and power as required.
5. Maintains a stabilized approach and recommended approach airspeed, or in its absence, not more than 1.3 V_{SO}, ±5 knots, with wind gust factor applied.
6. Makes smooth, timely, and correct control application during the roundout and touchdown.
7. Selects the proper landing path, contacts the water at the minimum safe airspeed with the proper pitch attitude for the surface conditions (AMES).
8. Touches down smoothly at minimum control airspeed (AMEL).
9. Touches down at or within 100 feet (30 meters) beyond a specified point, with no side drift, minimum float, and with the airplane's longitudinal axis aligned with and over the runway center/landing path.
10. Maintains crosswind correction and directional control throughout the approach and landing sequence.
11. Applies brakes (AMEL) or elevator control (AMES), as necessary, to stop in the shortest distance consistent with safety.
12. Completes appropriate checklist.

E. TASK: GLASSY WATER TAKEOFF AND CLIMB (AMES)

NOTE: If a glassy water condition does not exist, the applicant shall be evaluated by simulating the TASK.

REFERENCES: FAA-H-8083-3; POH/AFM.

Objective. To determine that the applicant:

1. Exhibits knowledge of the elements related to glassy water takeoff and climb.
2. Positions the flight controls and flaps for the existing conditions.
3. Clears the area; selects an appropriate takeoff path considering surface hazards and/or vessels and surface conditions.
4. Retracts the water rudders as appropriate; advances the throttle smoothly to takeoff power.
5. Establishes and maintains an appropriate planing attitude, directional control, and corrects for porpoising, skipping, and increases in water drag.
6. Utilizes appropriate techniques to lift seaplane from the water considering surface conditions.
7. Establishes proper attitude/airspeed, and accelerates to V_Y, ± 5 knots during the climb.
8. Retracts the landing gear, if appropriate, and flaps after a positive rate of climb is established.
9. Maintains takeoff power and V_Y ±5 knots to a safe maneuvering altitude.
10. Maintains directional control and proper wind-drift correction throughout takeoff and climb.
11. Completes the appropriate checklist.

F. TASK: GLASSY WATER APPROACH AND LANDING (AMES)

NOTE: If a glassy water condition does not exist, the applicant shall be evaluated by simulating the TASK.

REFERENCES: FAA-H-8083-3; POH/AFM.

Objective. To determine that the applicant:

1. Exhibits knowledge of the elements related to glassy water approach and landing.
2. Adequately surveys the intended landing area.
3. Considers the wind conditions, water depth, hazards, surrounding terrain, and other watercraft.

4. Selects the most suitable approach path and touchdown area.
5. Establishes the recommended approach and landing configuration and airspeed, and adjusts pitch attitude and power as required.
6. Maintains a stabilized approach and the recommended approach airspeed, ± 5 knots and maintains a touchdown pitch attitude and descent rate from the last altitude reference until touchdown.
7. Makes smooth, timely, and correct power and control adjustments to maintain proper pitch attitude and rate of descent to touchdown.
8. Contacts the water in the proper pitch attitude, and slows to idle taxi speed.
9. Maintains crosswind correction and directional control throughout the approach and landing sequence.
10. Completes the appropriate checklist.

G. TASK: ROUGH WATER TAKEOFF AND CLIMB (AMES)

NOTE: If a rough water condition does not exist, the applicant shall be evaluated by simulating the TASK.

REFERENCES: FAA-H-8083-3; POH/AFM.

Objective. To determine that the applicant:

1. Exhibits knowledge of the elements related to rough water takeoff and climb.
2. Positions the flight controls and flaps for the existing conditions.
3. Clears the area; selects an appropriate takeoff path considering wind, swells surface hazards, and/or vessels.
4. Retracts the water rudders as appropriate; advances the throttle smoothly to takeoff power.
5. Establishes and maintains an appropriate planing attitude, directional control, and corrects for porpoising, skipping, or excessive bouncing.
6. Lifts off at minimum airspeed and accelerates to V_Y, ± 5 knots before leaving ground effect.
7. Retracts the landing gear, if appropriate, and flaps after a positive rate of climb is established.
8. Maintains takeoff power and V_Y ±5 knots to a safe maneuvering altitude.
9. Maintains directional control and proper wind-drift correction throughout takeoff and climb.
10. Completes the appropriate checklist.

H. TASK: ROUGH WATER APPROACH AND LANDING (AMES)

NOTE: If a rough water condition does not exist, the applicant shall be evaluated by simulating the TASK.

REFERENCES: FAA-H-8083-3; POH/AFM.

Objective. To determine that the applicant:

1. Exhibits knowledge of the elements related to rough water approach and landing.
2. Adequately surveys the intended landing area.
3. Considers the wind conditions, water, depth, hazards, surrounding terrain, and other watercraft.
4. Selects the most suitable approach path, and touchdown area.
5. Establishes the recommended approach and landing configuration and airspeed, and adjusts pitch attitude and power as required.
6. Maintains a stabilized approach and the recommended approach airspeed, or in its absence not more than 1.3 V_{SO} $\pm$ 5 knots with wind gust factor applied.
7. Makes smooth, timely, and correct power and control application during the roundout and touch down.
8. Contacts the water in the proper pitch attitude, and at the proper airspeed, considering the type of rough water.
9. Maintains crosswind correction and directional control throughout the approach and landing sequence.
10. Completes the appropriate checklist.

I. TASK: GO-AROUND/REJECTED LANDING (AMEL and AMES)

REFERENCES: FAA-H-8083-3; POH/AFM.

Objective. To determine that the applicant:

1. Exhibits knowledge of the elements related to a goaround/rejected landing.
2. Makes a timely decision to discontinue the approach to landing.
3. Applies takeoff power immediately and transitions to climb pitch attitude for V_Y and maintains V_Y ±5 knots.
4. Retracts flaps, as appropriate.
5. Retracts the landing gear if appropriate after a positive rate of climb is established.
6. Maneuvers to the side of runway/landing area to clear and avoid conflicting traffic.
7. Maintains takeoff power and V_Y ±5 knots to a safe maneuvering altitude.
8. Maintains directional control and proper wind-drift correction throughout the climb.
9. Completes the appropriate checklist.

V. AREA OF OPERATION: PERFORMANCE MANEUVER

TASK: STEEP TURNS (AMEL and AMES)

REFERENCES: FAA-H-8083-3; POH/AFM.

Objective. To determine that the applicant:

1. Exhibits knowledge of the elements related to steep turns.
2. Establishes the manufacturer's recommended airspeed or if one is not stated, a safe airspeed not to exceed V_A.
3. Rolls into a coordinated 360° steep turn with at least a 50° bank, followed by a 360° steep turn in the opposite direction.
4. Divides attention between airplane control and orientation.
5. Maintains the entry altitude, ±100 feet (30 meters), airspeed, ±10 knots, bank, ±5°; and rolls out on the entry heading, ±10°.

VI. AREA OF OPERATION: NAVIGATION

A. TASK: PILOTAGE AND DEAD RECKONING (AMEL and AMES)

REFERENCE: AC 61-23/FAA-H-8083-25.

Objective. To determine that the applicant:

1. Exhibits knowledge of the elements related to pilotage and dead reckoning.
2. Follows the preplanned course by reference to landmarks.
3. Identifies landmarks by relating surface features to chart symbols.
4. Navigates by means of precomputed headings, groundspeed, and elapsed time.
5. Corrects for and records differences between preflight groundspeed and heading calculations and those determined en route.
6. Verifies the airplane's position within two (2) nautical miles of flight planned route.
7. Arrives at the en route checkpoints within three (3) minutes of the initial or revised ETA and provides a destination estimate.
8. Maintains appropriate altitude, ±100 feet (30 meters), and heading, ±10°.

B. TASK: NAVIGATION SYSTEMS AND RADAR SERVICES (AMEL and AMES)

REFERENCES: FAA-H-8083-3, AC 61-23/FAA-H-8083-25; Navigation Equipment Operation Manuals, AIM.

Objective. To determine that the applicant:

1. Exhibits knowledge of the elements related to navigation systems and radar services.
2. Demonstrates the ability to use an airborne electronic navigation system.
3. Locates the airplane's position using the navigation system.
4. Intercepts and tracks a given course, radial, or bearing, as appropriate.
5. Recognizes and describes the indication of station passage, if appropriate.
6. Recognizes signal loss and takes appropriate action.
7. Uses proper communication procedures when utilizing radar services.
8. Maintains the appropriate altitude, ±100 feet (30 meters) and heading, ±10°.

C. TASK: DIVERSION (AMEL and AMES)

REFERENCES: FAA-H-8083-25; AIM.

Objective. To determine that the applicant:

1. Exhibits knowledge of the elements related to diversion.
2. Selects an appropriate alternate airport and route.
3. Makes an accurate estimate of heading, groundspeed, arrival time, and fuel consumption to the alternate airport.
4. Maintains the appropriate altitude, ±100 feet (30 meters), and heading, ±10°.

D. TASK: LOST PROCEDURES (AMEL and AMES)

REFERENCES: FAA-H-8083-25; AIM.

Objective. To determine that the applicant:

1. Exhibits knowledge of the elements related to lost procedures.
2. Selects an appropriate course of action.
3. Maintains an appropriate heading and climbs, if necessary.
4. Identifies prominent landmarks.
5. Uses navigation systems/facilities and/or contacts an ATC facility for assistance, as appropriate.

VII. AREA OF OPERATION: SLOW FLIGHT AND STALLS

A. TASK: MANEUVERING DURING SLOW FLIGHT (AMEL and AMES)

REFERENCES: FAA-H-8083-3; POH/AFM.

Objective. To determine that the applicant:

1. Exhibits knowledge of the elements related to maneuvering during slow flight.
2. Selects an entry altitude that will allow the task to be completed no lower than 3,000 feet (920 meters) AGL.
3. Establishes and maintains an airspeed at which any further increase in angle of attack, increase in load factor, or reduction in power, would result in an immediate stall.
4. Accomplishes coordinated straight-and-level flight, turns, climbs, and descents with landing gear and flap configurations specified by the examiner.
5. Divides attention between airplane control and orientation.
6. Maintains the specified altitude, ±50 feet (15 meters); specified heading, ±10°; airspeed +5/−0 knots, and specified angle of bank, ±5°.

B. TASK: POWER-OFF STALLS (AMEL and AMES)

REFERENCES: FAA-H-8083-3, AC 61-67; POH/AFM.

Objective. To determine that the applicant:

1. Exhibits knowledge of the elements related to power-off stalls.
2. Selects an entry altitude that allows the task to be completed no lower than 3,000 feet (460 meters) AGL.
3. Establishes a stabilized descent in the approach or landing configuration, as specified by the examiner.
4. Transitions smoothly from the approach or landing attitude to a pitch attitude that will induce a stall.
5. Maintains a specified heading ±10°, in straight flight; maintains a specified angle of bank, not to exceed 20°, ±5°, in turning flight while inducing the stall.
6. Recognizes and recovers promptly as the stall occurs by simultaneously reducing the angle of attack, increasing power to maximum allowable, and leveling the wings to return to a straightand-level flight attitude with a minimum loss of altitude appropriate for the airplane.

7. Retracts the flaps to the recommended setting, retracts the landing gear, if retractable, after a positive rate of climb is established.
8. Accelerates to V_X or V_Y speed before the final flap retraction; returns to the altitude, heading, and airspeed specified by the examiner.

C. TASK: POWER-ON STALLS (AMEL and AMES)

REFERENCES: FAA-H-8083-3, AC 61-67; POH/AFM.

NOTE: In some high performance airplanes, the power setting may have to be reduced below the practical test standards guideline power setting to prevent excessively high pitch attitudes (greater than 30° nose up).

Objective. To determine that the applicant:

1. Exhibits knowledge of the elements related to power-on stalls.
2. Selects an entry altitude that allows the task to be completed no lower than 3,000 feet (920 meters) AGL.
3. Establishes the takeoff or departure configuration. Sets power to no less than 65 percent available power.
4. Transitions smoothly from the takeoff or departure attitude to a pitch attitude that will induce a stall.
5. Maintains a specified heading ±5°, in straight flight; maintains a specified angle of bank, not to exceed a 20°, ±10° in turning flight, while inducing the stall.
6. Recognizes and recovers promptly as the stall occurs by simultaneously reducing the angle of attack, increasing power to maximum allowable, and leveling the wings to return to a straightand-level flight attitude, with a minimum loss of altitude appropriate for the airplane.
7. Retracts flaps to the recommended setting, retracts the landing gear if retractable, after a positive rate of climb is established.
8. Accelerates to V_X or V_Y speed before the final flap retraction; returns to the altitude, heading, and airspeed specified by the examiner.

D. TASK: SPIN AWARENESS (AMEL and AMES)

REFERENCES: FAA-H-8083-3, AC 61-67; POH/AFM.

Objective. To determine that the applicant:

1. exhibits knowledge of the elements related to spin awareness by explaining:
2. Aerodynamic factors related to spins.
3. Flight situations where unintentional spins may occur.
4. Procedures for recovery from unintentional spins.

VIII. AREA OF OPERATION: EMERGENCY OPERATIONS

NOTE: Examiners shall select an entry altitude that will allow the single engine demonstrations TASK to be completed no lower than 3,000 feet (920 meters) AGL or the manufacturer's recommended altitude, whichever is higher. At altitudes lower than 3,000 feet (920 meters) AGL, engine failure shall be simulated by reducing throttle to idle and then establishing zero thrust.

A. TASK: EMERGENCY DESCENT (AMEL and AMES)

REFERENCES: FAA-H-8083-3; POH/AFM.

Objective. To determine that the applicant:

1. Exhibits knowledge of the elements related to an emergency descent.
2. Recognizes situations, such as depressurization, cockpit smoke and/or fire that require an emergency descent.
3. Establishes the appropriate airspeed and configuration for the emergency descent.
4. Exhibits orientation, division of attention, and proper planning.
5. Maintains positive load factors during the descent.
6. Completes appropriate checklists.

B. TASK: ENGINE FAILURE DURING TAKEOFF BEFORE V_{MC} (SIMULATED) (AMEL and AMES)

REFERENCES: FAA-H-8083-3; POH/AFM.

NOTE: Engine failure (simulated) shall be accomplished before reaching 50 percent of the calculated V_{MC}.

Objective. To determine that the applicant:

1. Exhibits knowledge of the elements related to the procedure used for engine failure during takeoff prior to reaching V_{MC}.
2. Closes the throttles smoothly and promptly when simulated engine failure occurs.
3. Maintains directional control and applies brakes (AMEL) or flight controls (AMES), as necessary.

C. TASK: ENGINE FAILURE AFTER LIFT-OFF (SIMULATED) (AMEL and AMES)

REFERENCES: FAA-H-8083-3; POH/AFM.

Objective. To determine that the applicant:

1. Exhibits knowledge of the elements related to the procedure used for engine failure after lift-off.
2. Recognizes a simulated engine failure promptly, maintains control, and utilizes appropriate emergency procedures.
3. Reduces drag, identifies and verifies the inoperative engine after simulated engine failure.
4. Simulates feathering the propeller on the inoperative engine. Examiner shall then establish zero-thrust on the inoperative engine.
5. Establishes V_{YSE}; If obstructions are present, establishes V_{XSE} or +5 knots, whichever is greater, until obstructions are V_{MC} cleared. Then transitions to V_{YSE}.
6. Banks toward the operating engine as required for best performance.
7. Monitors operating engine and makes adjustments, as necessary.
8. Recognizes the airplane's performance capabilities. If a climb is not possible at V_{YSE}, maintain V_{YSE} and return to the departure airport for landing, or initiates an approach to the most suitable landing area available.
9. Secures the (simulated) inoperative engine.
10. Maintains heading, ±10°, and airspeed, ±5 knots.
11. Completes appropriate emergency checklist.

D. TASK: APPROACH AND LANDING WITH AN INOPERATIVE ENGINE (SIMULATED) (AMEL and AMES)

REFERENCES: FAA-H-8083-3; POH/AFM.

Objective. To determine that the applicant:

1. Exhibits knowledge of the elements related to an approach and landing with an engine inoperative to include engine failure on final approach.
2. Recognizes engine failure and takes appropriate action, maintains control, and utilizes recommended emergency procedures.
3. Banks toward the operating engine, as required, for best performance.
4. Monitors the operating engine and makes adjustments as necessary.
5. Maintains the recommended approach airspeed ±5 knots, and landing

configuration with a stabilized approach, until landing is assured.

6. Makes smooth, timely and correct control applications during roundout and touchdown.
7. Touches down on the first one-third of available runway, with no drift and the airplane's longitudinal axis aligned with and over the runway center/landing path.
8. Maintains crosswind correction and directional control throughout the approach and landing sequence.
9. Completes appropriate checklists.

E. TASK: SYSTEMS AND EQUIPMENT MALFUNCTIONS (AMEL and AMES)

REFERENCES: FAA-H-8083-3; POH/AFM.

Objective. To determine that the applicant:

1. Exhibits knowledge of the elements related to systems and equipment malfunctions appropriate to the airplane provided for the practical test.
2. Analyzes the situation and takes appropriate action for simulated emergencies appropriate to the airplane provided for the practical test for at least five (5) of the following—
   a. partial or complete power loss.
   b. engine roughness or overheat.
   c. carburetor or induction icing.
   d. loss of oil pressure.
   e. fuel starvation.
   f. electrical malfunction.
   g. vacuum/pressure, and associated flight instruments malfunction.
   h. pitot/static.
   i. landing gear or flap malfunction.
   j. inoperative trim.
   k. inadvertant door or window opening.
   l. structural icing.
   m. smoke/fire/engine compartment fire.
   n. any other emergency appropriate to the airplane.
3. Follows the appropriate checklist or procedure.

F. TASK: EMERGENCY EQUIPMENT AND SURVIVAL GEAR (ASEL and ASES)

REFERENCES: FAA-H-8083-3; POH/AFM.

Objective. To determine that the applicant:

Exhibits knowledge of the elements related to emergency equipment and survival gear appropriate to the airplane and environment encountered during flight. Identifies appropriate equipment that should be aboard the airplane.

IX. AREA OF OPERATION: HIGH ALTITUDE OPERATIONS

A. TASK: SUPPLEMENTAL OXYGEN (AMEL and AMES)

REFERENCES: 14 CFR part 91; FAA-H-8083-3, AC 61-107; AIM, POH/AFM.

Objective. To determine that the applicant exhibits knowledge of the elements related to supplemental oxygen by explaining:

1. Supplemental oxygen requirements for flight crew and passengers when operating non-pressurized airplanes.
2. Identification and differences between "aviators" breathing oxygen" and other types.
3. Operational characteristics of continuous flow, demand, and pressure-demand oxygen systems.

B. TASK: PRESSURIZATION (AMEL and AMES)

REFERENCES: FAA-H-8083-3, AC 61-107; AIM, POH/AFM.

Objective. To determine that the applicant:

1. Exhibits knowledge of the elements related to pressurization by explaining—
   a. fundamental concept of cabin pressurization.
   b. supplemental oxygen requirements when operating airplanes with pressurized cabins.
   c. physiological hazards associated with high altitude flight and decompression.

NOTE: Element 2 applies only if the airplane provided for the practical test is equipped for pressurized flight operations.

2. Operates the pressurization system properly, and reacts appropriately to simulated pressurization malfunctions.

X. AREA OF OPERATION: MULTIENGINE OPERATIONS

NOTE: If the applicant is instrument rated, and has previously demonstrated instrument proficiency in a multiengine airplane or does not hold an instrument rating airplane, TASKs D and C need not be accomplished.

A. TASK: MANEUVERING WITH ONE ENGINE INOPERATIVE (AMEL and AMES)

REFERENCES: FAA-H-8083-3; POH/AFM.

NOTE: The feathering of one propeller shall be demonstrated in flight, in a multiengine airplane equipped with propellers which can be safely feathered and unfeathered. The maneuver shall be performed at altitudes and positions where safe landings on established airports can be readily accomplished. In the event a propeller cannot be unfeathered during the practical test, it shall be treated as an emergency.

Objective. To determine that the applicant:

1. Exhibits knowledge of the elements related to maneuvering with one engine inoperative.
2. Recognizes engine failure and maintains control.
3. Sets the engine controls, reduces drag, identifies and verifies the inoperative engine, and feathers appropriate propeller.
4. Establishes and maintains a bank toward the operating engine as required for best performance in straight and level flight.
5. Follows the prescribed checklists to verify procedures for securing the inoperative engine.
6. Monitors the operating engine and makes necessary adjustments.
7. Demonstrates coordinated flight with one engine inoperative (propeller feathered).
8. Restarts the inoperative engine using appropriate restart procedures.
9. Maintains altitude ±100 feet (30 meters) or minimum sink as appropriate and heading ±10°.
10. Completes the appropriate checklists.

B. TASK: V_{MC} DEMONSTRATION (AMEL and AMES)

REFERENCES: FAA-H-8083-3; POH/AFM.

NOTE #1 An applicant seeking an airplane—multiengine land (AMEL) rating, "Limited to Center Thrust," is not required to be evaluated on this TASK.

NOTE #2 Airplanes with normally aspirated engines will lose power as altitude increases because of the reduced density of the air entering the induction system of the engine. This loss of power will result in a V_{MC} lower than the stall speed at higher altitudes. Therefore, recovery should be made at the first indication of loss of directional control, stall warning, or buffet.

Do not perform this maneuver by increasing the pitch attitude to a high angle with both engines operating and then reducing power on the critical engine. This technique is hazardous and may result in loss of airplane control.

Objective. To determine that the applicant:

1. Exhibits knowledge of the elements related to V_{MC} by explaining the causes of loss of directional control at airspeeds less than , the factors affecting V_{MC}, and safe recovery procedures. V_{MC}
2. Configures the airplane at V_{SSE}/V_{YSE}, as appropriate—
   a. Landing gear retracted.
   b. Flaps set for takeoff.
   c. Cowl flaps set for takeoff.
   d. Trim set for takeoff.
   e. Propellers set for high RPM.
   f. Power on critical engine reduced to idle.
   g. Power on operating engine set to takeoff or maximum available power.
3. Establishes a single-engine climb attitude with the airspeed at approximately 10 knots above V_{SSE} or V_{YSE}, as appropriate.
4. Establishes a bank toward the operating engine, as required for best performance and controllability.
5. Increases the pitch attitude slowly to reduce the airspeed at approximately 1 knot per second while applying rudder pressure to maintain directional control until full rudder is applied.
6. Recognizes indications of loss of directional control, stall warning or buffet.
7. Recovers promptly by simultaneously reducing power sufficiently on the operating engine while decreasing the angle of attack as necessary to regain airspeed and directional control. Recovery SHOULD NOT be attempted by increasing the power on the simulated failed engine.
8. Recovers within 20° of the entry heading.
9. Advances power smoothly on operating engine and accelerates to V_{XSE}/V_{YSE}, as appropriate, ± 5 knots, during the recovery.

C. TASK: ENGINE FAILURE DURING FLIGHT (By Reference to Instruments) (AMEL and AMES)

REFERENCES: 14 CFR part 61; FAA-H-8083-3, FAA-H-8083-15.

Objective. To determine that the applicant:

1. Exhibits knowledge of the elements by explaining the procedures used during instrument flight with one engine inoperative.
2. Recognizes engine failure, sets the engine controls, reduces drag, identifies, and verifies the inoperative engine and simulates feathering appropriate engine propeller.
3. Establishes and maintains a bank toward the operating engine as required for best performance in straight and level.
4. Follows the prescribed checklists to verify procedures for securing the inoperative engine.
5. Monitors the operating engine and makes necessary adjustments.
6. Demonstrates coordinated flight with one engine inoperative.
7. Maintains altitude ±100 feet (30 meters), or minimum sink as appropriate and heading ±10°, bank ±5°, and levels off from climbs and descents within ± 100 feet (30 meters).

D. TASK: INSTRUMENT APPROACH—ONE ENGINE INOPERATIVE (By Reference to Instruments) (AMEL and AMES)

REFERENCES: 14 CFR part 61; FAA-H-8083-3, AC 61-27; FAA-S8081-4.

Objective. To determine that the applicant:

1. Exhibits knowledge of the elements by explaining the procedures used during a published instrument approach with one engine inoperative.
2. Recognizes engine failure, sets the engine controls, reduces drag, identifies and verifies the inoperative engine, and simulates feathering appropriate engine propeller.
3. Establishes and maintains a bank toward the operating engine, as required, for best performance in straight and level.
4. Follows the prescribed checklists to verify procedures for securing the inoperative engine.
5. Monitors the operating engine and makes necessary adjustments.
6. Requests and receives an actual or a simulated ATC clearance for an instrument approach.
7. Follows the actual or a simulated ATC clearance for an instrument approach.

8. Maintains altitude within 100 feet (30 meters), the airspeed within ±10 knots if within the aircraft's capability, and heading ±10°.

9. Establishes a rate of descent that will ensure arrival at the MDA or DH/DA, with the airplane in a position from which a descent to a landing, on the intended runway can be made, either straight in or circling as appropriate.

10. On final approach segment, no more than three-quarter-scale deflection of the CDI/glide slope indicator. For RMI or ADF indicators, within 10° of the course.

11. Avoids loss of aircraft control, or attempted flight contrary to the engine-inoperative operating limitations of the aircraft.

12. Complies with the published criteria for the aircraft approach category when circling.

13. Completes landing and appropriate checklists.

XI. AREA OF OPERATION: POSTFLIGHT PROCEDURES

NOTE: The examiner shall select TASK A and for AMES applicants at least one other TASK.

A. TASK: AFTER LANDING, PARKING, AND SECURING (AMEL and AMES)

REFERENCES: FAA-H-8083-3; POH/AFM.

Objective. To determine that the applicant:

1. Exhibits knowledge of the elements related to after landing, parking and securing procedures.
2. Maintains directional control after touchdown while decelerating to an appropriate speed.
3. Observes runway hold lines and other surface control markings and lighting.
4. Parks in an appropriate area, considering the safety of nearby persons and property.
5. Follows the appropriate procedure for engine shutdown.
6. Completes the appropriate checklist.
7. Conducts an appropriate postflight inspection and secures the aircraft.

B. TASK: ANCHORING (AMES)

REFERENCES: FAA-H-8083-3; POH/AFM.

Objective. To determine that the applicant:

1. Exhibits knowledge of the elements related to anchoring.
2. Selects a suitable area for anchoring, considering seaplane movement, water depth, tide, wind, and weather changes.
3. Uses an adequate number of anchors and lines of sufficient strength and length to ensure the seaplane's security.

C. TASK: DOCKING AND MOORING (AMES)

REFERENCES: FAA-H-8083-3; POH/AFM.

Objective. To determine that the applicant:

1. Exhibits knowledge of the elements related to docking and mooring.
2. Approaches the dock or mooring buoy in the proper direction considering speed, hazards, wind, and water current.
3. Ensures seaplane security.

D. TASK: RAMPING/BEACHING (AMES)

REFERENCES: FAA-H-8083-3; POH/AFM.

Objective. To determine that the applicant:

1. Exhibits knowledge of the elements related to ramping/beaching.
2. Approaches the ramp/beach considering persons and property, in the proper attitude and direction, at a safe speed, considering water depth, tide, current, and wind.
3. Ramps/beaches and secures the seaplane in a manner that will protect it from the harmful effect of wind, waves, and changes in water level.

APPENDIX 2

TASK VS. SIMULATION DEVICE CREDIT

Multiengine Land (MEL)

APPENDIX 2

AIRPLANE MULTIENGINE LAND TASK VS. SIMULATION DEVICE CREDIT

Examiners conducting the Commercial Pilot–Airplane Practical Tests with flight simulation devices should consult appropriate documentation to ensure that the device has been approved for training, testing, or checking. The documentation for each device should reflect that the following activities have occurred:

1. The device must be evaluated, determined to meet the appropriate standards, and assigned the appropriate qualification level by the National Simulator Program Manager. The device must continue to meet qualification standards through continuing evaluations as outlined in the appropriate advisory circular (AC). For airplane flight training devices (FTDs), AC 120-45 (as amended), Airplane Flight Training Device Qualifications, will be used. For simulators, AC 120-40 (as amended), Airplane Simulator Qualification, will be used.

2. The FAA must approve the device for training, testing, and checking the specific flight TASKs listed in this appendix.

3. The device must continue to support the level of student or applicant performance required by the PTS.

NOTE: Users of the following chart are cautioned that use of the chart alone is incomplete. The description and objective of each TASK as listed in the body of the PTS, including all notes, must also be incorporated for accurate simulation device use.

USE OF CHART

X Creditable

A Creditable if appropriate systems are installed and operating.

* Asterisk items require use of FTD or Simulator visual reference.

NOTES: 1. Use of Level 2 or Level 3 FTDs authorized only for those airplanes not requiring a type rating.

2. For practical tests, not more than 50 % of the maneuvers may be accomplished in an FTD or simulator UNLESS:

   a. each maneuver has been satisfactorily accomplished for an instructor, in the appropriate airplane, not less than three (3) times,

OR

   b. the applicant has logged not less than 500 hours of flight time as a pilot in airplanes.

3. Not all AREAS OF OPERATION (AOO) and TASKs required by this PTS are listed in the appendix. The remaining AOO and TASKs must be accomplished in an airplane.

4. Standards for and use of Level 1 FTDs have not been determined.

APPENDIX 2

FLIGHT TASK Areas f perftion:	AIRPLANE MULTIENGINE LAND FLIGHT SIMULATION DEVICE LEVEL 1	2	3	4	5	6	7	A	B	C	D
II. Preflight Procedures											
A. Preflight Inspection (Cockpit Only)		A	X	A	A	X	X	X	X	X	X
B. Cockpit Management		A	X	A	A	X	X	X	X	X	X
C. Engine Starting		A	X	A	A	X	X	X	X	X	X
D. Taxiing										X	X
E. Before Takeoff Check		A	X	A	A	X	X	X	X	X	X
IV. Takeoffs, Landings, and Go-Arounds											
A. Normal and Crosswind Takeoff and Climb										X	X
B. Normal and Crosswind Approach and Landing										X	X
C. Short-Field Takeoff and Climb								X	X	X	X
D. Short-Field Approach and Landing										X	X
E. Go-Around *			X			X	X	X	X	X	X
V. Performance Maneuver											
Steep Turns			X			X	X	X	X	X	X
VI. Navigation *											
B. Navigation Systems and ATC Radar Services		A			A	X	X	X	X	X	X
C. Diversion		A	X		A	X	X	X	X	X	X
D. Lost Procedures		A	X		A	X	X	X	X	X	X
VII. Slow Flight and Stalls											
A. Maneuvering During Slow Flight			X			X	X	X	X	X	X
VIII. Emergency Operations											
A. Emergency Descent			X			X	X	X	X	X	X
B. Maneuvering with One Engine Inoperative								X	X	X	X
C. Engine Inoperative—Loss of Directional Control Demonstration								X	X	X	X
D. Engine Failure During Takeoff Before V_{MC}								X	X	X	X
E. Engine Failure After Lift-Off (Simulated)								X	X	X	X
F. Approach and Landing with an Inoperative Engine (Simulated)											
G. Systems and Equipment Malfunctions		A	X	A	A	X	X	X	X	X	X
IX. Multiengine Operations											
A. Engine Failure During Flight (By reference to instruments)			X			X	X	X	X	X	X
B. Instrument Approach - All Engines Operating (By reference to instruments)		A	X		A	X	X	X	X	X	X
C. Instrument Approach - One Engine Inoperative (By reference to instruments)								X	X	X	X
X. High Altitude Operations											
B. Pressurization		A	X	A	A	X	X	X	X	X	X
XI. Postflight Procedures											
A. After Landing		A	X	A	A	X	X	X	X	X	X

Made in the USA
Monee, IL
04 November 2026

67377467R00066